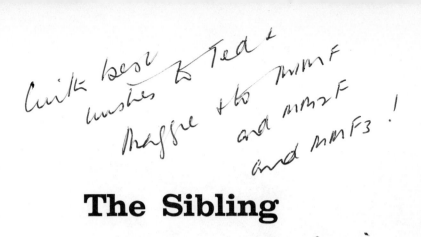

The Sibling

The Sibling

Brian Sutton-Smith, MM2
Teachers College, Columbia University

B. G. Rosenberg, MMFM4M
Bowling Green State University

HOLT, RINEHART AND WINSTON, INC.

New York   Chicago   San Francisco   Atlanta
Dallas   Montreal   Toronto   London   Sydney

Cover photograph by Ingbet

Library of Congress Catalog Card Number: 73–107671
SBN: 03–079055–7
Printed in the United States of America
1 2 3 4 5 6 7 8 9

*Dedicated to our older siblings who
will undoubtedly regard this as just
another form of harassment*

Preface

Most of the work for this study was carried out between the years 1962 and 1967, when both authors were members of the Psychology Department at Bowling Green State University in Ohio. We received support throughout from the Public Health Service.* Prior to 1962, the university authorities gave much special consideration through the Scholarly Advancement Committee and in other ways. Among the many colleagues who assisted us, knowingly or otherwise, in informal discussions were John T. Greene, Robert M. Guion, William W. Lambert, and John M. Roberts. And among the many others, including students who assisted us quite knowingly but more laboriously, were Judy Griffiths, Yael Orbach, Gael Goetchens, Elizabeth Greene, Anita Sharples, Regina Cryan, Lucy Gilbert, Stephen Goldman, and Roy Goldman.

Over the years much of our work would have been impossible without the support of the Bowling Green School authorities. Harlan Lehtomaa of Kenwood School, Bowling Green, put up with one or more of our studies every year for approximately a decade. Without his most

* MH 07994–01, –02, –03, –04, –05, –06, –07, –08; and MH 15786–01, –02, –03.

amiable acceptance of these annual irritations, we would have been, like most other psychologists, confined to the ambivalent fate of working solely with female undergraduates. We mention also the support of Gordon Schofield, Headmaster at Maumee Valley Country Day School, who welcomed our machinations and permitted us to extend the range of our inquiries to a higher socioeconomic group. Finally, we thank the Institute for Human Development, University of California at Berkeley, for granting us access to longitudinal data from the Guidance Study.

January 1970   BRIAN SUTTON-SMITH
                                                        B. G. ROSENBERG

Contents

CHAPTER

1

the sibling romance

Common sense dictates that our brothers and sisters, our siblings, have some effect upon our personality and development. What this effect may be is not always obvious. Being a sibling is not like being a male or female. There are no clear physiological signs that announce the biological character and expected actions of the person who is a firstborn, an older brother, or a younger sister. Sibling status is a *silent* variable, one that goes unannounced to both parties in a relationship, and yet it is like psychological maleness and femaleness in being the focus for systematic patterns of behavior. Sibling status has only recently become such a silent variable.

In European law, siblingship was linked in quite recent times to the preservation of intact properties by inheritance customs in which the individuals of a given sibling status received an allocation of property at the parents' death. In primogeniture, the firstborn male received all, or more rarely, in ultimogeniture, all went to the last born, and even more rarely, in secundogeniture or tertiogeniture, the property went to the second or third born. The corporate character of modern society, which renders sibling status unimportant for the distribution of property,

1

leaves unanswered the question of what other, less obvious types of inheritance may still be served by sibling status. Although modern society is not marked by those clear kinship divisions (along sex or generational lines) which were so important in structuring the function of earlier societies, we may well ask whether some informal, psychological "kin-ship" structures may not still be in service; whether there are not perhaps systems of attitudes between generations and between siblings which contribute in some way to a family's emotional viability. This would be the case if, for example, firstborn were more nurturing, like their parents, and really could carry on some of the family load of looking after younger ones. Alternatively, it might be argued that the current interest in sibling differences has appeared precisely because such differences are ceasing to be functional; without primogeniture to sustain sibling differences they are fading away, and one male in the family is becoming as good as another in matters of inheritance. In that case the studies in this area would be capturing history rather than human nature. Still, at this point, we wish to stress merely that sibling status is less overtly functional than it used to be, and to raise the question whether or not it may still serve some implicit functional ends.

Thus, while it would be conventional academic psychology to empha-size the importance of what we are doing by raising hopes that our information will contribute to a predictive science of siblings, it is per-haps premature to do so. Until now most of the work on siblings has attempted to show how parents make siblings different. The major point to be made in this book, however, is that siblings also make each other different. Whether the differences that are produced by other siblings, as well as by parents, enter permanently into their repertoires is much harder to say, although there is some evidence that they do. It seems saner to argue in the main that although siblings do have an influence upon each óther, whether or not that influence is enduring depends upon many other determinants. A primary research undertaking must be to understand the patterns of immediate influence and the character of the interactions and responses that result from them.

This book is an outgrowth of the rapid acceleration of systematic research carried out with siblings during the past decade, usually under the title of studies of ordinal position or of birth order. Ordinal position refers only to birth order (firstborn, second born, and so on), but as it has been established that the sex status of the sibling (whether boy or girl) is also of importance, we have chosen to use the term *sibling status* to refer to both of these characteristics in combination, birth order and sex.

One problem for the investigator working in this area is that there have been a number of popular studies that, while sometimes interesting and even exciting, are nevertheless presystematic in quality. They make many assertions on the basis of personal or clinical experience but do not present sufficient evidence to prove or disprove the theses they set forth.

A related problem is that the theme of sibling conflict as a struggle for power recurs throughout literary history. Whether power is truly the most important matter in the relationships of siblings is difficult to say, but it appears to be the most important consideration in traditional literature. Folk literature, which abounds with these power struggles, tends to favor the non-firstborn. Cain, the firstborn, suffers for despatching Abel. Jacob, the second born, prospers by taking the birthright from red and hairy firstborn Esau. Again, Joseph is the later born who, though much put upon by older brothers, eventually triumphs. This pattern apparently has many parallels in non-Western folklore and mythology (Herskovits & Herskovits, 1958). Adlerian psychology has suggested that in every fairy tale the youngest child surpasses all his brothers and sisters (Ansbacher & Ansbacher, 1956, p. 380). An analysis of the "winners" in 112 of Grimm's fairy tales about children (Scharl edition, 1948) showed, however, that while later born won most often, they did not always win. In 46 percent of the stories there was only one child, and in 71 percent of those, that child was the winner; that is, the outcome favored him or her. In 25 percent of the stories, there were two children. The outcome was favorable to both 36 percent of the time, to the firstborn 18 percent of the time, and to the second born 46 percent of the time. In 20 percent of the stories, there were three children; the first and second children won only 8 percent of the time, but the third child won 92 percent of the time! Four percent of the stories had seven children; one or more of the first six born won 33 percent of the time, and the seventh born won 66 percent of the time. Five percent of the stories had *many* children; the youngest won 53 percent of the time.

We incidentally noted that in most stories about large families, the siblings were of the same sex. If there is indeed more conflict among same-sex sibs (and the evidence seems to suggest that there is), it might be argued that the stories were of a mainly compensatory value for the later born. Of this import is Harris' speculation that fairy tales were mainly composed by the non-firstborn to meet their needs for immediate wish fulfillments (1964). On the other hand, it is sociologically possible that the last born actually had a favored status in large families (and there is some evidence for this also). At least one study shows that in instances of a large age gap between the last born and the preceding sibling, the last born's responses are, in general, more similar to those of only children than they are to the responses of either firstborn or later-born children (Miller & Zimbardo, 1965). Thus it is also conceivable that the tales were more a representation of the way things were than a compensation (Rosen, 1961).

Although folklore and tradition, for one reason or another, have given a larger place to the later born, the first scholarly studies of birth order in our own time associated the firstborn with superior achievement. A series of studies, beginning with Galton's *English Men of Science*, in 1874, have shown that firstborn exceed both non-firstborn and chance expectation in being Fellows of the Royal Society; in being sufficiently eminent to appear in the *Dictionary of National Biography*; in being scientists, professors, men of letters, Rhodes scholars; in being listed in *Who's Who*; in becoming eminent scientists; in having IQ's in the top 1 percent; and, more recently, in being finalists in National Merit Scholars competitions. The details of the studies underlying this impressive list can be found in a recent (albeit disputed) survey (Altus, 1966a). In the various groups, firstborn tended to be overrepresented by 10 to 30 percent more than their proportionate numbers in the population at large. A variety of reasons have been given for their preeminence, and these will be taken up later. At this point, all that need be noted is that the arguments continue as to whether firstborn really are superior, the fires being fed by a minority of studies showing contrary effects.

Studies of the relationship between eminence, scholarship, and sibling position were the first systematic studies of birth order, and they have outnumbered other studies of birth-order problems. Second in number of studies were various forms of problem behavior, in particular, delinquency, alcoholism, insanity, and physical disability. Unlike the studies of eminence, however, these studies have produced inconsistent results.

Turning from these early attempts at systematic research—some of which will be discussed in greater depth in later chapters—mention needs to be made of a number of writers who have added a great deal to the vitality of this field, even if their contributions have been disputable. Most important of these was Alfred Adler, who was the first to make a number of explicit claims about the effects of being born in different sibling positions and also the first to use sibling differences for explanatory purposes. He is worth quoting extensively because his ideas have become a part of the lore of family studies. Although Adler's statements have a categorical ring to them, he made it clear that none of the effects needed to occur. "It does not matter what really has happened, whether an individual is really inferior or not. What is important is his interpretation" (Adler, 1959, p. 124). The following statements about firstborn and second born and the boy with sisters will be considered in the chapters that follow:

> The oldest child also has well-defined characteristics. For one thing he has the advantage of an excellent position for the development of his psychic life. History recognizes that the oldest son has had a particularly favorable position. Among many peoples, in many classes, this advantageous status has become traditional. . . . Even where this tradition has not

actually become crystallized, as in simple bourgeois or proletarian families, the oldest child is usually the one who is accredited enough power and common sense to be the helper or foreman of his parents. . . . We can imagine that his thought processes are somewhat like this: "You are the larger, the stronger, the older, and therefore you must also be cleverer than the others." . . . If his development in this direction goes on without disturbance, then we shall find him with the traits of a guardian of law and order. . . . It is not surprising that such individuals are markedly conservative. . . . The striving for power in the case of a second-born child also has its especial nuance. Second-born children are constantly under steam, striving for superiority under pressure. . . . The fact that there is someone ahead of him who has already gained power is a strong stimulus for the second born.

Various combinations are possible. . . . The situation of an only boy among several girls is a case in point. A feminine influence dominates such a household, and the boy is pushed into the background, particularly if he is the youngest, and sees himself opposed by a closed phalanx of women. His striving for recognition encounters great difficulties. . . . A lasting insecurity, an inability to evaluate himself as a human being, is his most characteristic trait [Adler, 1959, pp. 125–128].

The most immediate empirical effect of Adler's work was a series of studies of sibling rivalry and, in particular, of the jealousy of older siblings (Levy, 1937). In one study such jealousy was said to be present in about 50 percent of the firstborn. It was more likely to occur if the parents disciplined the children inconsistently, if the mother was over-solicitous, and if the age gap between the siblings was between one-and-a-half and three years (Sewell, 1930). More jealousy was reported for same-sex siblings (particularly girls) than for opposite-sex siblings (Smalley, 1930). A practical outcome of these studies was that in subsequent years child-training experts began to be concerned with jealousy in children. Spock (1957) said:

Do your best to avoid jealousy. Jealousy is a strong emotion, even in grown-ups, but it is particularly disturbing to the young child just before the age of five. Such traits as selfishness and self-consciousness can often be traced back to the bitter jealousy created in the small child by the arrival of a baby brother or sister. . . . To prevent it or to minimize it is worth a lot of effort [p. 260].

He made a number of practical suggestions as to how this could be done.

One could speculate that a group of people who had been reading Spock for the past 20 years might produce fewer than 50 percent of firstborn who showed sibling rivalry. If true, we would have an interesting example of the way in which social science—or in this case what passes for it in child training literature—affects the populations that it studies by its own results and the practical recommendations flowing from them in such a way that they change. What begins as a normative law thus becomes an historical condition.

In some ways it seems unfair to mention Bossard and Boll's work in this chapter rather than with the more systematic studies that follow. They must be mentioned here, however, because in much of their work on siblings, their approach was only loosely systematic. Their data were derived from questionnaires, nondirective interviews, and family life-history documents written mainly by female informants. What is most evident in Bossard and Boll's richly descriptive data is a considerable feeling for the intensities of sibling interaction. Here are some of Bossard and Boll's reflections on sibling relationships:

> Their inclusive character . . . has, for example, a time aspect. Children in the same family may play together, work together, be together for long periods of time—for a day, week, and a year, etc. [1960, p. 89]. Sibling quarreling, etc. goes on between two personalities who are forced to spend many long hours together. The second aspect of the inclusiveness of the sibling relationship is the range of contacts included. For example, bathing, sleeping, playing, changing clothes, arguing, etc. [p. 90].
>
> Next to an intimate nature is the stark frankness of the sibling relationship. It is not one of company manners with doting parents hovering near to smooth out tangles and irregularities. Relationships between siblings permit little or no dissembling. In the terms of the baseball world, each solves the delivery of the other early in the game. No tricks will suffice, no deception will work. Siblings come to know each other by the book. They come to live largely with each other—to use the vernacular again—"with their hair down." Life among siblings is like living in the nude, psychologically speaking. Siblings serve as a constant crude awakening. On the other hand, siblings save each other from being with their parents and other adults too much. The significance of this is that they are kept from the unnatural environment, which the adult furnishes [p. 91]. Parents and other adults are less satisfactory companions for children than are other children because children treat each other more like equals [p. 92].

While most of the material in this work is not quantified, Bossard and Boll do report what looks like a significant difference between first-born not being favorable to living in large families and last born being favorable (a conclusion reverberated in Hall and Barger, 1964). Bossard, too, shows a penchant for sibling discipline as opposed to adult discipline. Perhaps there are echoes here of Bossard's own status as an only child. As we shall see later, only children perceive themselves as more powerless in relation to parents than do children with siblings. For example, Bossard and Boll say:

> Siblings feel that they understand each other and each other's problems and that often they do so better than the parents. . . . Siblings have a

better judgment often as to what constitutes misbehavior. Adults judge child behavior by adult standards; children judge it by child standards. This means that the discipline imposed by children on each other will seem more reasonable and have more meaning [1960, pp. 154–155].

Bossard deals at length with the extent to which children meet each other's needs for affection and act as counselors, leaders, and protectors, and also with the way in which older siblings serve as substitutes for parents in many families. This is especially true, he says, when there are many children and their birth extends over a long period of time. Many older children, he says, are exploited by being called upon to aid in the rearing of younger children to an extent that interferes with their own life plans. As a result, the older child tends to develop marked behavior trends in the direction of early maturing and of habits of responsibility and service to others. We have quoted Bossard and Boll at considerable length because their advocacy of siblings is a unique cause.

Another presystematic investigator is Walter Toman, who, in his book *The Family Constellation: Theory and Practice of a Psychological Game* (1961), presents the major and intriguing idea that the closer new relationships duplicate those of one's sibling childhood the more success- ful they will be. In particular, marriage relationships will be more suc- cessful if they duplicate earlier sibling relationships. Firstborn should marry second born, and opposite-sex siblings should have a better op- portunity for success in marriage than same-sex siblings. Unfortunately, the evidence presented is no match for the intriguing quality of the idea, and subsequent research has given only partial support. For example, Kemper (1966) found no tendency for husbands (256 business executives) to choose as marital partners those of the correct ordinal category, but he did show that some sibling dyads (pairs) made for a more satisfactory marriage self-rating than others. Thus, brothers with younger sisters (M1F) who married sisters with older brothers (MF2) reported a more satisfactory marriage relationship than brothers with older sisters (FM2) who married sisters with younger brothers (F1M).* Kemper suggested that this was because the former sibling relationship was more con- sistent with conventional expectations that the male will be dominant. As we shall see later, however, the former (M1F and MF2) was also a more satisfactory sibling dyad during development, for both of the offspring, than was the latter (F1M and FM2). This fact may say something about the ways in which the influence of sibling relationships affect subsequent self-ratings rather than about compatibilities peculiar to marriage. Kem- per's results seemed to show also that marriages between an older and a younger sibling did better than those between two younger or two older siblings, at least as far as the husbands were concerned; he

---

* See Chapter 3 for explanation of symbols used in referring to sibling statuses.

suggested this was because marital power relationships were less uncertain and conflictive in the former than in the latter two categories. In another study of engaged and married couples which compared male-dominant with female-dominant dyads, Lu Yi-Chang (1952) showed that in the male-dominant couples the women were (more often than chance would lead one to expect) the only or the youngest siblings and in the female-dominant couples, the women were more often the oldest or middle-born siblings. There was no relationship, however, between dominance and the husband's sibling order.

The final work to which we wish to call attention in this introduction is *The Promised Seed: A Comparative Study of Eminent First and Later Sons* by Irving Harris (1964). Harris recorded the ordinal positions of famous historical figures and, by placing them into firstborn and later-born groups, attempted to see how they differed as politicians, poets, philosophers, and military men. Regardless of the success of his speculations, it is interesting to play the game of comparisons that Table 1.1, drawn from his study, permits.

TABLE 1.1 Famous Firstborn and Later Born

	First	Later
Psychologists	Freud, Jung	Adler, Rank
Philosophers	W. James, Kant	Dewey, Hume, Bacon
Social philosophers	Montesquieu, Marx, Engels, Martin Luther King	Rousseau, Machiavelli, Ghandi, Hobbes, Lenin, Voltaire
Mathematicians	Pascal, Newton, Einstein	Descartes
Literary	Shakespeare, Shaw, Carlyle, Nietzsche, Spengler	Dostoevski, Camus, Tolstoy, Lawrence, H. James
Military	Grant, Patton, Pershing, Washington	Lee, Napoleon, Bismarck, Rommel
Musicians	Beethoven	Wagner
Poets	Milton, Pope, Schiller, Shelley, Dante	Wordsworth, Poe, Coleridge
Politicians	T. Roosevelt, Truman, Perry, Caesar, Churchill, Frederick the Great, Alexander the Great, Mussolini, Hitler	Jackson, Cesare Borgia, Richelieu, Cromwell, Trotsky, John Kennedy, Khrushchev

SOURCE: Based on data from Irving Harris, *The Promised Seed: A Comparative Study of Eminent First and Later Sons* (Glencoe, Ill.: Free Press, 1964).

Harris has many things to say about the differences between his firstborn and later-born famous men, but his most comprehensive view is that the former favor a synthesizing conceptual approach to reality and the latter an analytic approach. Contrasts between the birth orders that can be drawn from his various discussions are shown in Table 1.2.

TABLE 1.2 Comparison of Characteristics of Firstborn and Later Born

Firstborn	Later Born
Holistic	Atomistic
Humanistic	Skeptical
Contemplative	Activistic
Romantic	Realistic
Epic	Lyric
Depressive	Cheerful
Internalizing	Externalizing
Messianic	Propagandistic
Conceptual	Perceptual
Abstract	Concrete
Stable	Histrionic

SOURCE: Based on data from Irving Harris, *The Promised Seed: A Comparative Study of Eminent First and Later Sons* (Glencoe, Ill.: Free Press, 1964).

Among Harris' many interesting contrasts is one between Freud and Adler. According to Harris, Freud as a firstborn naturally focused his attention on the special and intimate relationship of the child to his parents, and used the notion of the Oedipal attachment to explain many diverse phenomena such as dreams, familial affairs, sociology, and mythology. Adler, by contrast a fourth-born son less close to his parents, emphasized the importance of siblings and concentrated on the singular effects of the drive for status and power and upon the importance of compensating for feelings of inferiority. In Freudian therapy one comes to terms with an inner moral authority. In Adlerian therapy, one adapts to the external world and its moral power. In Freud, one is occupied in dealing with the internal voices of the parents (superego versus ego versus id); in Adler, the unified self deals with the external world. Harris suggests that there is in these two psychologies a classic difference between the internalized psychology of one who began life with power and with a special access to parental figures, and the externalized psychology of one who began life with relatively little power and little direct access to parents. In a more recent, unpublished work entitled *Violence and Responsibility: The Politics of Identity,* Harris extends his discussion of firstborn and second born to suggest that the firstborn's lifelong and special access to parents gives him a certain self-righteousness. Under attack, he reacts with a stiffening of personal identity; the later born is more prepared to shift ground, and to change one identity for another. The firstborn is more consistent and rigid about his principles; the second born more pragmatic and deceptive. For these reasons, Harris asserts, firstborn make more dangerous political leaders. They are most likely to "fight than switch." Harris' most striking piece of evidence (perhaps coincidental) is the finding that of 35 presidents, 7 of the 20 firstborn have led the country into war (Madison, Polk, Lincoln, Wilson, Roosevelt, Truman, and Johnson), but only one of the later sons did (McKinley).

Following Harris, we might speculate that it is natural for Irving Harris, a firstborn, to have such a comprehensive and synthesizing conceptual approach to this field. Whereas it is just as natural for two later born, like the present authors, to see things in a more fragmentary light, and like Adler, to emphasize the importance of child-child relationships. Too much of the research in this field, we might protest, has been carried on by firstborn academics!

These brief references to folklore, early scholarship, Adler, Bossard, Toman, and Harris must suffice to give the reader some of the "romantic" flavor attached to the study of sibling relationships. In the following chapter we move on to a brief statement of the methodological and theoretical considerations that must be taken into account before one can begin to be sure that the apparent evidence really does tell us something about siblings, and not covertly about social status, family size, or mother's age, or—as in the case of some of the writers mentioned above—about the particular experiences and sibling status of the writers themselves.

some
precautions

In 1931 Harold Jones, director of the Institute of Child Development at Berkeley, University of California, published a paper entitled "Order of Birth in Relation to the Development of the Child." This should have become the bible for future ordinal research. Judging by the relative dearth of such studies from 1933 until 1960, however, his major effect was to dispel enthusiasm for this subject matter. Up to that time practically all of the studies were of the *survey* type, in which large samples of varying ordinal position were assessed for differences in intelligence, rates of delinquency, and so forth. A later review by Murphy, Murphy, and Newcomb was of similar critical import (1937). Since 1960 there has been a resurgence of ordinal position studies, many of which are of an experimental character in which a subject's responses to systematically varied stimuli are recorded in controlled situations. Although these later experimental studies define the testing situation much more precisely than did the earlier studies in which subjects merely responded to inventories, intelligence tests, and so forth, many of the recent studies have been no more precise about the characteristics of the subjects being tested than were the earlier studies cited critically by Jones. In this recent literature we do not have, as clearly as we might wish, an example of the cumulative effects of science (Sampson, 1965; Warren, 1966).

The present work owes a great deal to Jones's criticisms. His article is a natural starting point for the systematic study of siblings (1931). In summary form, the methodological considerations that must be taken into account and which will be introduced at appropriate points in the following chapters involve: (1) *Ordinal position criteria:* the specific ordinal positions, the sex of the siblings, the age spacing between siblings, the duration of relationships. (2) *Developmental criteria:* the age of the subjects, the age of the mother, the fads in child rearing. (3) *Family criteria:* family size, family completeness. (4) *Cultural criteria:* cross-cultural contrasts, subcultural contrasts, socioeconomic variables and demographic variables.

While it is not possible to study siblings systematically without methodological controls, it is not possible to study them sensibly without adequate attention to underlying theoretical approaches. In Wittgenstein's terms, when about to play a language game, we must be as explicit as possible about the syntax we intend to use.

First, siblings are a part of a family, so that usually some notions about the nature of *interactions within the family* are assumed by an investigator, even if he does not make them explicit. In the present book the assumption is made that it is possible to talk about siblings and family in terms of three basic classes of interaction: those between the children themselves (which are discussed in Chapters 3 and 4), those between parents and children (Chapters 5, 6, and 7), and those involving child-child and parent-child interactions at the same time (Chapter 8).

Second, in discussing the way in which family members *influence* each other, a number of types of behavior must be taken into account. Parents shape child behavior by their reactions to it. They also provide models for the children's own learning. Similarly it may be supposed that siblings shape each other's behavior through their own reactions, and in addition model after each other's examples. Siblings may also shape the behavior of their parents, causing parents, in turn, to react toward them in different ways. Again, some behavior may be a reaction to the behavior of others (as in anger) without copying it, and sometimes without influencing it. In sum, in discussing influence in families it is necessary to keep in mind the paradigms of operant learning and modeling, and also to allow for those responses which are often evoked by other family members, even when they have neither been asked for nor exemplified—responses which will be termed here *counteractive.* But these complicated matters are best discussed in the context of the data they are supposed to illuminate in the chapters that follow. They are introduced here to highlight the important point that a book of this character is not ultimately concerned with sibling status. It is concerned with discovering those conditions of learning that have created the effects regularly associated with sibling status. As many have said, sibling status is a locus for specific learning contingencies, and once these contingencies have been formulated the study of sibling status itself becomes irrelevant. It should be added, however, that at the present time this is a rather optimistic orientation.

A work of this type must be concerned also with a number of the classic problems of developmental psychology. For example: Are sibling differences caused by the differential treatment of the parents only during early childhood, or are these differences sustained by a continuing difference in parental care? Do the observed differences remain the same throughout development, or are there important changes in these differences as a function both of varying life stage contingencies and of transformations in the subjects themselves? Again: Are the responses dealt with largely the results of genetics, learning, or disequilibrium?

The position will be taken throughout this work that without careful methodological controls it is often difficult to know whether apparent differences between siblings are really due to sibling status, or are actually a product of family size or socioeconomic status or some other nonsibling variable. Again, in reviewing sibling differences that seem reliable, the focus will be on interactions between siblings themselves and between siblings and parents. With social responses such as the sibling interactions dealt with in this book, it makes sense to us that our major concern will be with the types of influence and development usually handled within a social learning framework.

CHAPTER 3

sex status effects

SPECIFIC SIBLING STATUS

Child to child interactions have not been a major concern in sibling research, but they have been the focus for the present investigators. One step toward the more adequate understanding of such interactions is to specify exactly the particular sibling statuses being considered and, therefore, the nature of the sibling dyad involved in these interactions (older brother with younger brother, older brother with younger sister, and so on). Traditionally comparisons were made only between firstborn and second born without considering the specific ordinal position and the sex of the sibling. And yet, as Jones has shown, those comparisons were subject to various types of error (1931). The real pregnancy order (given infant mortality) might be very different from the observed birth order, so that if one did wish to claim some superiority or difference for one of the orders, then it would be necessary first to check whether, say, the existing oldest in the family was really the firstborn, or whether the real firstborn had disappeared in miscarriage, stillbirth, or some other prenatal disorder. In one early estimate cited by Jones, it was suggested that at least 15 percent of alleged firstborn were actually second born, and 15 percent of alleged second born were

14

actually third born. Reduced infant mortality could be expected to make this a decreasingly important distinction. Again, unless a comparison of first- and second-born subjects is controlled for the presence of physical defect, there will normally be more defects among the firstborn subjects, with a consequent effect on the psychological measures being taken. Firstborn are more susceptible to both paranatal (at time of birth) and prematurity disorders. They are also more susceptible to prenatal casualty and stillbirth. These factors mean that the firstborn that survive should show more defects, more sequelae, or after effects, of prematurity and yet at the same time be more highly selected than comparable second-born groups. No one seems to have sorted out adequately the statistical proportions of any firstborn sample that may be weighted in these different manners. Jones makes the judgment that "it would appear that from the standpoint both of physical injury and of possible after effects of birth trauma, the first born is in an unfavorable position" (1931, p. 214).

While there is not as yet any solid evidence on the matter, it also appears as if the birth experience of being firstborn rather than later born may possibly introduce physical variations that have later psychological effects. The suggestion has been made, for example, that the first child gives off antigens, substances to which the mother reacts by creating antibodies, which in turn create a minor shock for the second born, increasing his activity level. Whether or not the causal chain is as indicated here, in several studies, second born have been shown to be more active babies. On the other hand, the firstborn are, in general, smaller babies, growing at a faster rate and catching up the difference by the age of two years, at which time they surpass the second born in weight and height; at adulthood, however, there is no difference (Clausen, 1966). But these are qualifications meant to indicate that ordinal position is not always what it seems, when firstborn and others are compared. They bear the warning that if firstborn are physiologically distinctive, it is not impossible that much of the phenomena that are attributed to parental behavior in later chapters may actually be, at least in part, stimulated by the offspring themselves. What has been attributed to parental shaping behavior may be a description only of one part of a complex synchrony between infant and mother, each being both active or reactive in different ways over a period of time.

Helen Koch has shown that even if such factors are controlled, the broad category of firstborn or second born is not really a unified category psychologically (1954). There are, for example, various types of firstborn: there are those who are boys, those who are girls, those who have younger brothers, and those who have younger sisters. Koch, in her work, dealt with the 8 types of ordinal position in the two-child family. In similar fashion in some studies of our own, we have dealt with the 24 types of the three-child family. In this latter work there were 8 possible types of firstborn; 8, of second born; and 8, of third born. In four-child families there would be 16 types of firstborn! And in five-child families there would be 32 types of firstborn! It is clear both in Koch's results and our own that studies which merely contrast firstborn with later born mask a variety of differences between specific categories even when they do yield some differences. And as we shall show in later chapters, occasionally these particular sibling positions reverse an ordinal difference which otherwise holds for the larger groups of which they are a part. For example, although firstborn are generally more powerful than second born, younger brothers with older sisters (FM2) sometimes have more power than those older sisters (F1M) (see Chapter 4). One obvious reason for the paucity of studies to clarify the matter is that they take so long to carry out. For example, in one of our studies of children in the three-child family, it took us several years to collect a sufficient number of subjects for adequate statistical comparisons between each of the 24 positions ($N = 25$ per position). The percentage of people in such families is, after all, quite small. We should add, however, that some simplifications can be used which combine categories in various ways in order to overcome such sample deficits. Thus, in addition to firstborn versus second-born contrasts, like- versus opposite-sex siblings can be contrasted: siblings from families of various sizes in which all siblings are males can be contrasted with those in which all siblings are females, if family size is controlled. Again, subjects can be grouped according to immediately older and younger siblings and treated as members of three-child families.

For simplicity of reference to sibling statuses, throughout the pages that follow, we have used a system of symbols as shown in Table 3.1.

TABLE 3.1 Symbols Indicating Sibling Statuses

An only male:	M
An only female:	F
Boy with younger brother:	M1M
Boy with younger sister:	M1F
Boy with older brother:	MM2
Boy with older sister:	FM2
Girl with younger brother:	F1M
Girl with younger sister:	F1F
Girl with older brother:	MF2
Girl with older sister:	FF2

The same system holds for all other size families. The letter M refers to male; the letter F, to female. The order of sequence from left to right is the order of males and females in the family from oldest to youngest. The particular person we are referring to is the one with the number after his M or F; that number also represents his birth order in the family. Thus the 24 three-child family positions are labeled for males as:

M1MM, M1MF, M1FM, M1FF
MM2M, MM2F, FM2M, FM2F
MMM3, MFM3, FMM3, FFM3

and for females as:

F1FF, F1MF, F1FM, F1MM
FF2F, FF2M, MF2F, MF2M
FFF3, FMF3, MFF3, MMF3

Koch's study stands alone in the adequacy of its methodological controls of the specific ordinal positions in the two-child family. She used 384 six-year-old children—all from normal and complete families in the city of Chicago—and all matched for socioeconomic status, residence, and age. There were 48 subjects for each of her 8 two-child positions. Each of these groups was subdivided into three groups according to the age difference between the siblings, that is, under two years (0–2), between two and four years (2–4), or between four and six years (4–6). All her subjects were between five and six years of age. Similarly throughout the present study most of the data have been derived from members of the 8 two-child family sibling statuses, though the ages of the subjects were usually at preadolescence or college levels. In some cases Koch's age-spacing controls were used, but in most they were not. Comparisons generally involved subjects whose ages were within four years of each other. Given the present decrease in birth rate and the fact that, in any case, firstborn and second born usually comprise more than 50 percent of the population, it seems likely that the restriction of this study largely to two-child families has increasing rather than decreasing relevance.

The procedure in this chapter is to deal, first, with inventory measures of preference and interest which reflect direct interactive effects between siblings but could be regarded as only a rather superficial indication of these influences; second, with cognitive and observational measures which might be considered to indicate more enduring sibling effects. Showing that siblings, whether male or female, younger or older, affect each other in terms of their own characteristics does not control against the possibility that these differences are actually being brought about by differential parent attention. This would not, of course, detract from the empirical worth of discovering the differences that go with different types of sibling interaction, even though they were mediated by differential parent action. But it does present problems for our under-standing of sibling interaction itself. Although the studies reported in this chapter mainly deal with sibling-sibling interactions, and thus make the assumption that variations in patterns of sibling behavior are a direct outcome of the nature of the sibling statuses, it is only in several studies dealing with interactions involving both parents and children that we actually begin to control for parent-sibling influences as well as sibling-sibling influences.

SEX ROLE PREFERENCES

The present investigators used measures of sex role preference at both preadolescence and college levels. All these measures gave essentially the same verdict: namely that the preference patterns of the subjects were affected by the sex of their siblings. At the simplest level, all this means is that each subject has knowledge of the sort of choices that his sibling would make. Whether their behavior is also affected is another question (Vroegh, 1969). Table 3.2 shows the scores of a sample of small-town, northwestern Ohio children, in two- and three-child families on the authors' 180-item play and game scale (Sutton-Smith & Rosenberg, 1959). As had been expected, in the two-child families the boys with brothers had significantly higher masculinity scores, and the girls with sisters had significantly higher femininity scores (Rosenberg & Sutton-Smith, 1964b). Each sex is differentiated on its own sex scale, but not on the opposite-sex scale.

TABLE 3.2 Effects of Like- and Opposite-Sex Siblings

Sex of Siblings of Male Subjects	N	Masculinity	Femininity	Sex of Siblings of Female Subjects	N	Masculinity	Femininity
			Two-child Families				
Male	32	18.13*	4.34	Male	23	6.91	10.69
Female	47	15.28	4.45	Female	32	7.03	13.69*
			Three-child Families				
Male	19	14.95	3.37	Male	23	8.22	13.78
Female	18	18.83	4.47	Female	20	7.10	13.95

SOURCE: B. G. Rosenberg and B. Sutton-Smith, "Ordinal Position and Sex Role Identification," *Genetic Psychological Monographs*, 1964, **70**, 297–328(b).

* $p = < .05$. Comparisons are made between items adjacent to each other, vertically.

What these scores mean is indicated in Table 3.3, which shows the types of items on the masculinity and femininity scales. Boys with brothers more often said they played and liked the types of items in the masculinity scale than did boys with sisters; likewise girls with sisters more often marked the types of items in the femininity scale (Rosenberg & Sutton-Smith, 1959).

It should be noted that in the three-child families the effects were different. Boys with two sisters scored higher on the masculinity scale, not on the femininity scale as had been expected. This type of outcome, which we shall call a *counteractive phenomena*, requires some new explanation, to be given later.

TABLE 3.3 Game Choices Differentiating between Boys (N = 928) and Girls (N = 973)

Masculinity Game	Femininity Game
Bandits	Dolls
Soldiers	Dressing up
Cowboys	House
Cops and robbers	Store
Cars	School
Spacemen	Church
Marbles	Actors
Bows and arrows	Actresses
Throw snowballs	Stoop tag
Darts	Ring-around-the-Rosey
Wrestling	London Bridge
Baseball	Farmer in the Dell
Football	In and out the window
Basketball	Drop the handkerchief
Boxing	Mulberry Bush
Shooting	Hopscotch
Fish	Jump rope
Hunt	Jacks
Use tools	Mother, May-I?
Climbing	Dance
Make radios	Sewing
Model airplanes	Cooking
Toy trains	Knit
Work machines	Crochet
Build forts	Cartwheels

SOURCE: B. G. Rosenberg and B. Sutton-Smith, "The Measurement of Masculinity and Femininity in Children: An Extension and Revalidation," *Journal of Genetic Psychology*, 1964, **104**, 259–264.

NOTE: All items differentiate between the sexes beyond the < .01 level.

At the college level, using the masculinity-femininity scale of the Minnesota Multiphasic Personality Inventory (MMPI), with a sample of the members of two-child families controlled more carefully for socioeconomic status than with this preadolescent group, the present investigators again found the same type of difference between boys with brothers and boys with sisters, though there were no significant differences between girl groups. Table 3.4 gives the mean scores of the only-child and two-child families.

TABLE 3.4 Mf Scores from the MMPI
(N = 93 males; 160 females)

Males		Females	
FM2	67.37	F1F	45.16
M	63.19	F	46.81
M1F	62.47	F1M	46.82
M1M	61.52	FF2	47.61
MM2	60.25	MF2	48.33

SOURCE: Adapted from B. Sutton-Smith and B. G. Rosenberg, "Age Changes in the Effects of Ordinal Position on Sex Role Identification," *Journal of Genetic Psychology*, 1965, **107**, 61–73.

In Table 3.4 the higher the score the higher the femininity for males, but the lower the score the higher the femininity for females. In this and the following studies by the present investigators the samples were predominantly of upper-lower and lower-middle socioeconomic status, of rural or small-town origin in Ohio. The students were education students taking courses in psychology.

INTERESTS

There is not a great deal of data on sibling interest preferences. What data there are follow pretty much along the lines of the sex preference data presented above. As an example we present below a study of college student responses to the investigators' recreational inventory (Males = 83; Females = 90). The items of this inventory were grouped according to whether they were games of pure chance (C), chance and strategy (CS), pure strategy (S), physical skill and strategy (PS), pure physical skill (P), outdoor skill activities (O), social activities (So), or vicarious activities (V). The items in each of these categories are listed in Table 3.5. Mean responses for each ordinal position are given in Table 3.6. The significant findings are stated in Table 3.7.

On this inventory the effect of brothers on sisters was as noticeable as the effect of sisters on brothers. Both affected each other with their

TABLE 3.5 Recreation Inventory

<table>
<tr><td colspan="3" align="center">Sibling Interest Preferences</td></tr>
<tr><td>Chance (C)</td><td>Hockey</td><td>Acrobatics</td></tr>
<tr><td>Horse racing</td><td>Tennis</td><td>Surfing</td></tr>
<tr><td>Betting</td><td>Softball</td><td></td></tr>
<tr><td>Gambling</td><td>Ping-Pong</td><td>Social Activities (So)</td></tr>
<tr><td>Bingo</td><td>Billiards</td><td>Drill team</td></tr>
<tr><td>Craps</td><td>Snooker</td><td>Square dance</td></tr>
<tr><td>Roulette</td><td>Volleyball</td><td>Folk dance</td></tr>
<tr><td>One-armed bandit</td><td>Croquet</td><td>Ballet</td></tr>
<tr><td>Dice</td><td></td><td>Tap dancing</td></tr>
<tr><td>Coin toss</td><td>Physical Skill (P)</td><td>Band</td></tr>
<tr><td>Old Maid</td><td>Golf</td><td>Scouting</td></tr>
<tr><td></td><td>Miniature Golf</td><td>Debates</td></tr>
<tr><td>Chance and</td><td>Archery</td><td>Ballroom dancing</td></tr>
<tr><td>Strategy (CS)</td><td>Track</td><td>Public speaking</td></tr>
<tr><td>Canasta</td><td>Shuffleboard</td><td>Smoking</td></tr>
<tr><td>Euchre</td><td>Bowling</td><td>Dating</td></tr>
<tr><td>Poker</td><td>Quoits</td><td>Nesting</td></tr>
<tr><td>Solitaire</td><td>Horseshoes</td><td>Choir</td></tr>
<tr><td>Bridge</td><td>Darts</td><td>Dramatics</td></tr>
<tr><td>Pinochle</td><td>Tug o'war</td><td>Visiting</td></tr>
<tr><td>Hearts</td><td>Swimming races</td><td>Committee work</td></tr>
<tr><td>Rummy</td><td>Shooting matches</td><td>Student affairs</td></tr>
<tr><td>Cards</td><td>Gym contests</td><td>Parties</td></tr>
<tr><td>Monopoly</td><td></td><td>Coaching</td></tr>
<tr><td>Clue</td><td>Outdoor Skill (O)</td><td>Kissing games</td></tr>
<tr><td></td><td>Cycling</td><td>Club activities</td></tr>
<tr><td>Strategy (S)</td><td>Hunting</td><td>Fraternity acts</td></tr>
<tr><td>Chess</td><td>Climbing</td><td>Bull sessions</td></tr>
<tr><td>Scrabble</td><td>Swimming</td><td>Picnics</td></tr>
<tr><td>Checkers</td><td>Camping</td><td>Hayrides</td></tr>
<tr><td>Dominoes</td><td>Horseback riding</td><td>Majorettes</td></tr>
<tr><td>Draughts</td><td>Skiing</td><td>Cheer leading</td></tr>
<tr><td>Tic Tac Toe</td><td>Water skiing</td><td>Church activities</td></tr>
<tr><td>20 Questions</td><td>Roller skating</td><td>Telephoning</td></tr>
<tr><td>Puzzles</td><td>Ice skating</td><td>Charities</td></tr>
<tr><td></td><td>Fishing</td><td></td></tr>
<tr><td>Physical Skill and</td><td>Boating</td><td>Vicarious Activities (V)</td></tr>
<tr><td>Strategy (PS)</td><td>Hiking</td><td>Reading</td></tr>
<tr><td>Football</td><td>Sledding</td><td>Movies</td></tr>
<tr><td>Basketball</td><td>Tobogganing</td><td>Dining out</td></tr>
<tr><td>Boxing</td><td>Weight lifting</td><td>Watching sports</td></tr>
<tr><td>Wrestling</td><td>Exercising</td><td>Watching shows</td></tr>
<tr><td>Baseball</td><td>Car traveling</td><td>Watching TV</td></tr>
<tr><td>Handball</td><td>Airplane flying</td><td>Listening to music</td></tr>
<tr><td>Soccer</td><td>Parachute jumping</td><td>Daydreaming</td></tr>
</table>

TABLE 3.6 Mean Scores of Sibling Responses to Recreation Inventory

	N	C	CS	S	PS	P	O	So	V
M	18	0	3.47	.86	8.90	6.72	91.00	12.83	15.86
M1M	17	.17	3.22	.57	10.28	3.34	7.67	15.76	13.53
M1F	16	.03	2.00	.85	6.97	2.64	6.58	15.69	9.55
MM2	17	0	4.90	.27	14.76	4.40	12.02	15.55	22.17
FM2	15	.04	2.04	1.31	4.33	2.75	5.29	14.60	20.56
F	18	.05	3.27	.10	1.51	1.51	6.21	19.19	15.79
F1F	19	.06	2.28	.36	2.32	1.42	4.93	11.83	7.73
F1M	18	.01	3.70	.53	3.14	2.01	12.13	19.15	14.65
FF2	19	.02	3.17	.14	1.29	1.56	6.55	15.39	13.38
MF2	16	.13	6.45	.82	5.96	3.09	10.47	20.87	14.09
Total M	83	.05	3.13	.74	9.22	3.96	8.12	14.96	15.64
Total F	90	.05	3.59	.36	2.61	1.83	8.01	17.19	13.21

own interests. Boys with sisters showed fewer athletic interests and a directionally greater interest in strategy. Girls with brothers showed more athletic interest and were also more social in their interests. Altus also found girls with brothers to have more heterosexual interests (1967b). In the Koch study of playmates, opposite-sex siblings said they had more

TABLE 3.7 Ordinal Differences in Recreation

Game	Subjects	t	Level of Significance
Strategy	Males > females	2.15	<.05
Physical Skill and Strategy	Males > females	5.09	<.01
	Males with brothers > Males with sisters	2.02	<.05
	Females with brothers > Females with sisters	2.47	<.02
	MF2 > FF2	3.28	<.01
	MF2 > F1F	2.28	<.02
	MF2 > F	3.01	<.01
Physical Skill	Males > females	3.91	<.01
	Females with brothers > Females with sisters	2.16	<.05
Social	F1M > F1F	2.04	<.05
	F > F1F	2.05	<.05
Vicarious	MM2 > M1F	2.12	<.05
	F > F1F	2.29	<.05

opposite-sex playmates (1956c). The *Vicarious* category (watching TV, reading, and so forth) was highest for second-born males.

In another study involving interests, this time with the Strong Vocational Inventory, it was found that when same- and opposite-sex dyads were contrasted in terms of their vocational interest, the all-boys' dyad (M1M and MM2) preferred typical masculine entrepreneurial activities, the all-female dyad (F1F and FF2) preferred typical female secretarial-type activities, and the opposite-sex dyads showed a stronger interest in creative occupations such as artist, music performer, author, and architect (Sutton-Smith, Roberts, & Rosenberg, 1964). The greatest male interest in these occupations was shown by the boy with the younger sister (M1F), and the greatest female interest by the girl with the older brother (MF2). Next highest in level of interest in these creative occupations was the dyad of older sister and younger brother (F1M and FM2). As this latter dyad also had the highest scores on the clinical measure of conflict (MMPI) already mentioned, we have come to speak rather loosely of the former group as being a creative dyad (M1F and MF2), and the latter group as a creative-clinical dyad (F1M and FM2).

These effects of the sibling's sex on a subject's sex-role preferences, athletic, social, and occupational interests are fairly straightforward. That is, each sex increases the importance of its own sex role typical traits in the subject. In anticipating the following chapters on affiliation, conformity, and achievement, it might be expected also that girl siblings should heighten affiliation and conformity in subjects (as these are more typical female traits) and that boy siblings should heighten achievement in subjects (as these are more typical male traits). There is evidence for such trends, however, only in the case of the specific second-born sibling positions—FM2 and MF2. The former becomes more affiliative, and the latter more achieving.

The effect of opposite-sex siblings on creativity is more problematic, partly because creativity itself is a relatively novel variable in research. The finding of such an opposite-sex effect is certainly consonant with that literature which tends to show that the opposite-sex parent is a major influence in the lives of creative persons (Barron, 1963). It is consonant also with some recent research showing that children who score high on measures of creativity tend to have inconsistent sex role patterns (Singer, 1968). Following the present investigators' interest in seeking a source in the character of the interactions themselves, there is available information from a study in which we used the Bene-Anthony Family Relations Test with 7- to 11-year-old boys (Sutton-Smith, Rosenberg, & Houston, 1968). This study will be described in Chapter 7. Of greatest import in the present context was the finding that same-sex siblings showed more positive responses to and intake from siblings than did opposite-sex siblings ($p < .001$). Opposite-sex siblings (males) showed more concern with the self ($p < .05$). These results suggest that opposite-sex siblings decrease the amount of sibling social relations within the family and increase their self-concern. When it is realized that each sex in this situation has the problem of assimilating into its self-concern the opposite-sex influence of the other siblings—boys taking in the more domestic-aesthetic concerns typical of girls, and girls taking in the more athletic-vocational concerns typical of boys—then possibly the relative solitariness within the sibling relationships and the self-concern may suggest that what occurs is a greater degree of intraceptiveness in siblings in these opposite-sex families. The wider range of stimulation may cause a resort to intraceptive forms of resolving the opposite-sex influence. Possibly the general inhibition of direct physical contact between opposite-sex siblings is a further inducement to such intraception. Still, these are but speculations rather meagerly supported by the data. All we know at present is that there is apparently some interrelationship between opposite-sex role influence and creativity.

Koch's findings with 6-year-olds indicated that cross-sex siblings were both more stimulating and stressful for each other than same-sex siblings at the age of 6 years, though this effect was more marked upon the firstborn than the second born (1956c). This, in itself, was an interesting reversal insofar as the firstborn usually have a greater effect upon the second born than vice versa. The same-sex siblings (at least those between 0 and 2 years in age differences) seemed to have a relatively less stimulating effect upon each other. In her more recent study of twins (1966) Koch also found that monozygotic, or identical, twins were less stimulating to each other than dyzygotic, or non-identical, twins (Sutton-Smith, 1968). At the age of 6, at least, having to associate and deal with someone of a different sex-role identity was more challenging (stimulating and threatening) than having to deal with someone of the same identity. On Freudian grounds Koch suggests we might have expected that those of the same sex would have been more stimulating and had more to fight about, having the same opposite-sex parent as a goal object (1956c). The import of her data, however, seems to be that at age 6, the child's greater concern is with his own sex identity. This would perhaps be consistent with the Freudian position if it could be accepted that at age 6 the proposed oedipal crisis had been succeeded by attempts at appropriate role identification. It will suffice to note here that whichever way this matter is construed, the sex of the siblings is a key factor in their influence upon each other.

ABILITIES

Even if we grant that siblings have a considerable effect upon each other's interests and preferences, this may still be regarded as a fairly superficial matter. At least, the self-report methods of measurement (inventories and questionnaires) discussed to this point have been fairly superficial. It would give much more satisfaction to one skeptical about the importance of siblings to be shown that measures of ability varied in some systematic way with sibling status.

Four studies all point to an effect of brothers upon sisters. At the childhood level, Koch (1954) found siblings with brothers were superior to siblings with sisters on verbal meaning and quantitative tests, but findings were restricted to those with a 2- to 4-year-old age gap. Schoonover (1959), working with sibling pairs in early childhood also found that siblings with brothers were better on tests of language, literature, science, social studies, and arithmetic. Altus found a similar effect of male siblings on University of California students (1962). Firstborn students (either male or female) from two-child families earned higher scores on quantitative tests if they had male siblings. Rosenberg and Sutton-Smith, however, found a reverse effect with 900 subjects in the two-child family (female subjects with girl siblings were higher than female subjects with male siblings), but in the three-child family, male siblings had an elevating effect on the quantitative scores of second- and third-born girls, particularly if there were two males and if they were older (1964c; 1966). These contradictory results have in common that females are more affected by males than vice versa, but the other discrepancies remain unresolved.

The problem with these simple comparisons may be that they do not control adequately for age-spacing effects. While ordinal and sex effects were very evident in all of Koch's work, her most recurrent effect was due to age spacing (1955b). Most important, the 2–4 year age spacing seemed to heighten all other differences. When the gap was only 0–2 years, the siblings were more alike, which seemed to foster the girls' development but to have a somewhat depressive effect upon that of the boys. Children who were closer had more interests in common, played with each other more, were lonelier without each other, did not want to be rid of each other, and played with each other's friends more. When they were 4 to 6 years apart, however, the siblings had much less effect upon each other, a situation which was seemingly more stimulating to boys as judged by the fact that they were then more outgoing. It might be supposed that the close relationship of siblings fosters a mutual dependency which is more consonant with female sex role traits, and that the larger age gap fosters a mutual independence which is more consonant with male sex role traits.

As a test of this proposition and as a replication of Koch's age-spacing data, the present investigators contrasted age-spacing data for over 600 college-age females and over 300 college-age males in two-child families. It was argued that if the close spacing is facilitative for girls, then closely spaced girls would have higher college entrance scores; but if the large spacing is facilitative for boys, then boys at larger spacings should have higher scores (Rosenberg & Sutton-Smith, in press).

It was found that males' scores rose as the distance between siblings increased, and females' scores decreased with distance, although the contrast held best for the two- to three-year age gap where scores were low for males and high for the females. Closer examination of the specific ordinal positions also revealed that for girls high scores went with a small age gap if they had same-sex siblings, rather than opposite-sex siblings. This finding is consistent with the hypothesis that it is the sex role congruency of the closeness that facilitates their intellectual functioning. Being able to be dependent upon a sister older or younger by two to three years appears to have a facilitative effect, intellectually, upon girls. By contrast, boys' scores were not so much elevated by large distance as they were depressed at the same two- to three-year-age spacing. Here it was the second boys (FM2 and MM2) that appeared to suffer the most. One interpretation might be that boys at that close spacing suffer from the degree of competition involved. As neither argument holds for the one-year spacing, however, it needs to be acknowledged that the two- to three-year spacing must permit some contrast effect, which underlies these various interactions. Whatever the interpretation, these studies have borne out Koch's emphasis on the importance of age spacing in sibling studies as well as the fact that sibling status can have an effect not only on patterns of preference and interest but also on more enduring competences such as those measured by ability tests. We would mention as a caution that in a follow-up study with children rather than adults it has been found that these age-spacing effects also vary with the type of test being used (intelligence, reading, and the like) (Rosenberg, Goldman, & Sutton-Smith, 1969).

INTERACTION

To this point then it has been established that the sex of the sibling has a variety of effects on the subject. Tentatively it looks as if males and females both affect each other on measures of sex role preference and on interests, but that the results are more one-sided for the ability data, the more prevalent effect being that of males on female siblings. Although the consensuality across these various types of data is quite impressive, an investigator would have even more confidence if such differences between the siblings of like and opposite sex could actually be observed.

We present below, first, a summary of Koch's teachers' ratings of children's behavior, then some data from studies of our own in which children's interactions were directly observed (1955b). In her study of the 8 two-child family positions, Koch made use of teacher ratings. The raters, it should be noted, "were all women teachers, authority figures to our subjects, and the behavior samples on which the teachers' judgments of the children were based were chiefly school behaviors" (1955b, p. 17). The ratings themselves she grouped as having to do with attitudes toward adults (friendliness, affectionateness), social attitudes (friendliness to children, gregariousness, leadership, popularity), emotional attitudes (aggressiveness, kindness, excitability, intensity, moodiness), work attitudes (interests, enthusiasms, originality, playfulness, tenacity), and sissiness and tomboyishness.

Despite the biased effect of such ratings, Koch's work yielded some important consistencies across her various types of data (ratings by teachers, interviews, and projective tests with children). A major example is her study (1957) of the relationship between sibling status and the characteristics of the children chosen as playmates. Although most children chose playmates of the same age and sex, comparatively more older siblings said they preferred younger playmates, and younger siblings said they more often preferred older playmates. That is, there was a parallel between those they had as siblings and those they chose as playmates. Furthermore, opposite-sex siblings more often made opposite-sex playmate choices. Koch's teacher ratings permit a further check of these playmate choice differences. If opposite-sex siblings show a higher preference for play with opposite-sex playmates, it seems reasonable to suggest that they will show more of these opposite-sex characteristics in their own behavior. In a paper entitled "Sissiness and Tomboyishness in Relation to Sibling Characteristics" (1956d), Koch gives some slight evidence for this proposition on the basis of these two ratings (sissiness and tomboyishness) alone. In particular, she showed that the boy with an older sister (FM2) was more often rated a sissy than the boy with an older brother, and the girl with an older brother (MF2) was more often rated a tomboy than the girl with the older sister. But there were important age-spacing effects which complicated these relationships. Much stronger evidence in favor of the expected relationships was provided by Brim's reanalysis (1958) of the total body of Koch's rating data. Brim adopted the Parsonian distinctions between instrumental task roles (masculine) and expressive, socioemotional roles (feminine). Brim's judges were able to assign 30 of Koch's ratings to either the instrumental or expressive role category on theoretical grounds. "Some of the traits were stated in a negative way which made them, while pertinent to the role, incongruent with the role conception. Thus, 'uncooperativeness with the group' seemed clearly relevant to the expressive role but as an incongruent trait" (p. 7). Role traits were said to be either highly masculine if congruent (aggressiveness) and low masculine if incongruent (dawdling), or highly feminine if congruent (affectionateness), or low feminine if incongruent (anger). Koch's results were scored so that each group was compared with every other group on all 30 traits and scored as superior or inferior on each trait. The sibling group was then assigned a high or low rating on this trait, but only if this group was superior or inferior to all other same-sex groups on this comparison. Table 3.8 adapted from Brim (1958) shows the traits judged as fitting each category and the number of times each sex group was found to be extreme on that category. It is clear that the opposite-sex sibling overlaps the other sex traits to a greater extent than the same-sex siblings. The differences are most noticeable for the girls with male siblings. In sum, here is evidence of a broad-scale consistency between Koch's teacher ratings,

TABLE 3.8 The Variables in Brim's Reanalysis

High Masculine Rating	Low Masculine Rating	High Feminine Rating	Low Feminine Rating
1. Tenacity	1. Dawdling and procrastinating	1. Affectionate	1. Anger
2. Aggressiveness	2. Wavering in decision	2. Obedience (responds to sympathy)	2. Quarrelsome
3. Curiosity		3. Approval from adults	3. Revengeful-ness
4. Ambition		4. Speedy recovery from emotional upset	4. Teasing
5. Planfulness		5. Cheerfulness	5. Insistence on rights
6. Responsibility		6. Kindness	6. Exhibitionism
7. Originality		7. Friendliness to adults	7. Uncooperative with group
8. Competitiveness		8. Children	8. Upset by defeat
9. Self-confidence			9. Jealousy
			10. Negativism
			11. Tattling

		Number of Ratings on Which Each Group Was Superior		
F1F	5	15	33	16
FF2	7	18	36	14
	(12)	(33)	(69)	(30)
F1M	20	3	33	7
MF2	20	0	48	0
	(40)	(3)	(81)	(7)
M1M	9	12	0	41
MM2	12	13	10	33
	(21)	(25)	(10)	(33)
M1F	6	14	12	42
FM2	0	19	23	15
	(6)	(33)	(35)	(57)

SOURCE: Adapted from O. G. Brim, "Family Structure and Sex Role Learning by Children: A Further Analysis of Helen Koch's Data," *Sociometry,* 1958, **21,** 1–16.

the basic sibling positions, and the children's choice of playmates. The evidence shows that at age 6, one's siblings have a key effect on one's sex role attributes, at least as these are observed by teachers.

Having made this major point, however, it is important that we note that there are many complexities in this evidence. Thus, all-girl families were highest on both feminine congruent and incongruent traits, but they did not score much higher on these traits than did boys with sisters. Girls with brothers scored high only on congruent feminine and congruent masculine traits. They, therefore, presented an extremely healthy picture. In reverse, the boys with sisters scored highly on incongruent masculine and incongruent feminine traits. Even the all-boy families had their highest scores on the incongruent traits. These are strange results, and although they do show effects of opposite- versus same-sex siblings, they clearly have much to do with the age of the subjects, the traits being rated, and the fact that the ratings were done by female teachers. Brim points out that in making the ratings the teachers implicitly had rated the boys and girls on different scales: "The girl with an extreme masculine trait, extreme, that is, for a girl, receives a very high rating; a boy with the same absolute degree of such a trait or even more of it, would on the boys' scale, not be extreme and his rating would be consequently reduced. In the subsequent analysis of variance, where the male and female ratings are treated as if on the same absolute scale, certain girls, extremely high for girls, would score significantly higher than even certain boys high on the trait" (1958, p. 13).

In a study carried out by the present investigators with the assistance of Elizabeth Greene during 1967, differences were observed in school classrooms. Subjects were chosen from grades 1 to 6 in an elementary school in Bowling Green, Ohio. Children were included in the study who had siblings of the same or opposite sex (but not children who had mixed-sex siblings). In the two-child families there were 23 boys with opposite-sex siblings (M1F and FM2), and 16 with same-sex siblings (MM2 and M1M). There were 16 girls with opposite-sex siblings (F1M and MF2), and 20 girls with same-sex siblings (F1F and FF2). In three-child families there were five boys with all brothers, and eight boys with all sisters; there were seven girls with all brothers, and nine with all sisters. The subjects were observed in random order on four occasions in classroom situations. On each occasion the child was observed until he became involved in an interaction with a teacher or a classmate. In each case it was noted who initiated the interaction and, if a classmate, whether this was a boy or a girl. The number of observed interactions is indicated in Table 3.9.

In the two-child family, second born with opposite-sex siblings interact more with opposite-sex peers in the classroom than do second born with same-sex siblings. Firstborn are not thus affected by the sex of their

TABLE 3.9 Observed Interaction of Siblings

Subjects		Interaction observed with		
	Teacher	Same-Sex Peer	Opposite-Sex Peer	
Two-child Family				
Firstborn with same-sex siblings (M1M and F1F) N=15	6	37	17	ns
Firstborn with opposite-sex siblings (M1F and F1M) N=17	4	46	18	
Second born with same-sex siblings (MM2 and FF2) N=21	6	58	20	p < .01
Second born with opposite-sex siblings (FM2 and MF2) N=22	16	32	40	
Three-child Family				
All-female N=9	5	25	6	ns
Females with male siblings N=7	1	12	15	
All male N=5	1	12	7	ns
Males with female siblings N=8	8	16	8	

NOTE: ns = no significant statistical difference between these two groups. $p < .01$ = probability is less than one chance in a hundred that these two groups are *not* responding differently.

siblings. In the three-child family the same relationship tends to hold for females, though not significantly, but it does not hold for males. This is an interesting difference as it parallels that noted in Table 3.2, above, where boys with two sisters did *not* report higher femininity. That is, while boys with one sister in the two-child family respond as if affected by the sister, the boy with two sisters in the three-child family does not show their influence. Males also initiate relatively more interactions as compared to females. The numbers for each sibling group are shown in Table 3.10.

In sum, the evidence presented in this chapter implies that the sex character of a subject's sibling has effects on the subject's interests, preferences, abilities, and behavior.

TABLE 3.10 Interactions Initiated and Received

	Interaction Initiated	Interaction Received
Males with male siblings	44	20
Males with female siblings	55	37
Females with female siblings	41	39
Females with male siblings	26	38

THE CROSS-SEX EFFECT

There are findings in the literature which are perhaps most appropriately dealt with at this point, although they are somewhat anomalous in character. In a number of studies firstborn males and second-born females have made similar responses and have been dissimilar as a group from firstborn females and second-born males. Without going into great detail on these studies, Table 3.11 summarizes some of the contrasts that have been made between the two groups.

TABLE 3.11 Opposite-Sex Contrasts

	Firstborn Female Second-born Male	Firstborn Male Second-born Female
Koch, 1954		Better at tests of spatial relationships
Sampson, 1962	Less conforming	More conforming
Singer, 1964	More Machiavellian	Less Machiavellian
Schooler and Caudill, 1964		Higher rates of psychiatric hospitalization
Miller & Zimbardo, 1965	Less likely to cite need for distraction as a reason for affiliation	Cite distraction
	Less uncertain under conditions of threat	More uncertain
Smart, 1965	Less often members	More often members of social and recreational clubs
Haven, 1967	More cognitively complex— use more different concepts for describing associates and family members	Less conceptually complex
Eisenman, 1968a		Preference for more complex visual designs

In general, the responses on the left (firstborn female and second-born male) are more "masculine," or "tough minded," and those on the right (firstborn male and second-born female) are more "feminine," or "tender minded." Although we do not know which particular ordinal positions are causing these differences, there is a lead in Koch's discovery that at age 6, it is the F1F and the MM2 who together have more masculine characteristics than the M1M and FF2. The former are more popular (1955b, p. 39) and have more masculine characteristics like dawdling and procrastination (1956b, p. 295). Again in Chapter 6, it is MM2 and F1F who are the least conforming and FF2 and M1M (and M1F) who are the most conforming in the Bragge and Allen study (1966). In a study with anthropologist John M. Roberts (Sutton-Smith, Roberts, & Rosenberg, 1964), we likewise got a similar pairing of MM2 and F1F against M1M and FF2. The occupations which distinguished these two groups were as follows:

MM2: Physician, osteopath, dentist, veterinarian, farmer, policeman, mortician, physical director of YMCA.

M1M: Production manager, personnel director, public administrator, YMCA secretary, city school superintendent, social worker, certified public accountant, accountant, office man, purchasing agent, banker, sales manager, lawyer, president of a manufacturing concern.

F1F: Nurse, dentist, laboratory technician, physician, dietitian, physical therapist.

FF2: Lawyer, social worker, YWCA secretary.

These findings suggest that the opposite-sex effect is most likely to occur when same-sex siblings are being contrasted. In our own data given above, we argued that the occupations of the MM2 and F1F emphasize direct and physical modalities of control, while the occupations of the M1M and FF2 emphasize indirect and symbolic modalities of control. The polarization of physical and symbolic (or, as we called it, of potency versus strategy) is parallel to that between masculine and feminine. It is understandable on the basis of the arguments above that MM2 will be "physicalistic" and FF2 "social-symbolic" because these are both typical of their sex role.

There is a possible linkage with Schooler and Caudill's data on psychiatric hospitalization in which the later-born females were over-represented, and in upper-status groups, the firstborn males were over-represented (1964). While this mental illness literature is not notable for the consistency of discovered effects, Schooler, in a number of studies, does seem to have persistently documented a tendency for later-born females in Western culture and for firstborn males in Japanese culture to show higher rates of psychiatric hospitalization (1961, 1964). They argue that what is involved here is that where the male in Japan has a traditional firstborn status, the higher pressures upon him as an adult, after earlier indulgences as a child, make him more susceptible to stress. Alternately, the later-born female in Western culture may be considered to suffer relative parental neglect, and to be more susceptible to stress for that reason. While the collapse of primogeniture in Western culture may have removed the same degree of stress from firstborn males, Schooler and Caudill (1964) contend that there is still more evidence of it in upper- than in middle-status males (Barry & Barry, 1967).

As the converse of these sibling sex role arrangements, there is some evidence that sex role deviation (as contrasted with psychiatric deviation) is more likely to occur with the sibling positions on the left-hand side of Table 3.11. The evidence is too slight, however, to warrant more than passing attention as a possible focus for future study (Gundlach & Riess, 1967).

Perhaps the most important contribution to the series of studies showing these parallel cross-sex effects is Mary Rothbart's doctoral dissertation *Birth Order and Mother-Child Interaction* (1967), in which she compared children from two-child families of the same sex (M1M and MM2; F1F and FF2). Once again she found many contrasts between the M1M–FF2 and the F1F–MM2. The mother was observed interacting with her child while he worked on puzzles, and conversed with her, while he played. The mothers appeared to be milder and more praising in their treatment of the M1M–FF2, and more critical in their treatment of F1F–MM2. This treatment parallels nicely the Table 3.11 results, which may be read now to show that those who have been treated with more praise are more "tender minded," and those who have been treated more harshly are more "tough minded."

Rothbart's explanation has a psychoanalytic connotation, namely that the "mother felt a special attraction toward the first boy and a sense of rivalry toward the first girl, leaving the second boy with fewer expressions of approval than the first, and the second girl with more expressions of approval than the first." (1967, p. 86) This explanation can be complemented by another which has role-taking connotations, and parallels the argument presented with the age-spacing data above, namely that being born first implies closeness to parents and the carrying out of surrogate responsibilities. Both of these requirements are more typical of female sex role requirements and so facilitate the development of girls (F1F), but place stress on the firstborn boy (M1M). Being second born implies being more autonomous in modeling from various sources, a behavior more sex-role typical for males and, therefore, facilitative for boys (MM2) but less auspicious for girls (FF2).

In an interview study on children's power tactics carried out by these investigators during 1966 and cited in the next chapter, it was found that males more often said that the younger sibling had more fun, and females more often said that the older sibling had more fun, which is consistent with this explanation of sex-role typicality of these roles but should not occur if the parents' behavior was the sole determining factor. That is, Rothbart's mothers were giving the worst treatment to those who are said here to have more fun (F1F–MM2), and the best treatment to those not so chosen (M1M–FF2). Probably, in some way yet to be determined, the characteristics in Table 3.11 are an interactive product of these contrary conditions. The mothers may be tougher toward their firstborn daughters (F1F) and their second-born sons (MM2), but these two groups of subjects find support from the fact that the demands made upon them and the circumstances they suffer are congruent with sex role expectancies. Contrarily perhaps, the more sensitively treated M1M and FF2 find their relationships with their parents less propitious for dealing with the sex role expectancies of others.

It might be added that this is merely the first of a number of occasions throughout this work, when the contradictory data across sibling statuses appear to require the application of varying principles of influence in the present case, parental influence versus sex role expectancies.

COUNTERACTIVE SEX ROLE BEHAVIOR

Most of the discussion to this point has been in terms of the intensity of the direct effects of siblings upon each other. The notion has already been introduced, however, that in some cases the effect of one sex upon the other appears to cause a counteractive emphasis. Preadolescent boys with two sisters (as compared with one) did not give a higher feminine self-report. Boys in classrooms with two sisters (as compared with one) did not interact more with members of the opposite sex. An additional study by the present investigators adds more evidence concerning this effect. It was also unusual in containing all family members in the same sample. The sample was composed of 160 college sophomore females and 89 males, their actual siblings (males and females), and their mothers and fathers. The Gough Scale of Psychological Femininity (Fe) (1952) was administered directly to the subject's own sibling, mother, and father. All subjects were members of two-child families. While the many results are reported in detail elsewhere, the most interesting outcome for the present purposes is revealed by Table 3.12 (Rosenberg & Sutton-Smith, 1968). In this table the higher the score, the higher the acknowledged femininity on the part of the father respondents.

These results indicate that the more males there are in the family, the more feminine is the father's score; alternatively, the more females in the family, the more masculine is his score. That is, fathers of the

TABLE 3.12 Fathers' Femininity Scores as a Function of Sibling Constellation

In Various-sized Families	Scores
In all-girl families (F1F and FF2) (N = 90)	16.55
In boy-girl families reported by females (F1M and FM2) (N = 70)	17.46
In boy-girl families reported by males (M1F and FM2) (N = 49)	17.00
In all-male families (M1M and MM2) (N = 40)	18.35

SOURCE: Adapted from B. G. Rosenberg and B. Sutton-Smith, "Family Interaction Effects on Masculinity-Femininity," *Journal of Personality and Social Psychology*, 1968, 8, 117–120.

male dyads are relatively more willing, on the Gough scale, to admit to anxiety, discomfort, sensitivity, emotional disturbances, and so on, which are admissions about behaviors traditionally regarded as feminine. The difference is both significant and most striking between fathers of the all-girl dyad and fathers of the all-boy dyad. Lansky (1964) has also shown that such sibling patterns can have an effect on parents' femininity scores. He worked with 5-year-old children, so that we may assume these influences persist throughout development.

In the Rosenberg and Sutton-Smith study (1968) the mothers' femininity scores showed no such relationship to sibling composition. We can gain some understanding of this phenomenon by a study of the patterns of correlations of mothers, fathers, and two siblings in this study. In the all-girl group family, the girls' scores correlated with each other and with those of the mother, but the fathers' scores did not. The father's femininity score showed no significant correlations with any of his other family members. But in the boy-girl and the boy-boy dyad, by contrast, the scores of all family members intracorrelated in a variety of ways with each other (Cryan, 1968). From these various findings, it may be inferred that males with "too much feminine" influence may indeed have been resisting or counteracting it, or in some way in conflict with it. This impression is heightened by a reexamination of the observational study mentioned earlier. In this study, when the children were observed interacting with the same sex, opposite sex, or teacher, a two-minute observational protocol was also collected on each child. A content analysis of the items in this protocol yielded the data in Table 3.13, in which

TABLE 3.13 Observed Behavior in Classroom Interactions

Behavior	Males with Male Sibs (N = 38)	Females with Female Sibs (N = 35)	Males with Female Sibs (N = 43)	Females with Male Sibs (N = 32)
Impulsive (fidgets; inattentive; wanders; is disruptive)	8	6	24	13
Works at desk	12	12	9	6
Talks to teacher	4	10	1	4
Dutiful involvement	14	15	17	17

subjects with same-sex siblings were contrasted with subjects of opposite-sex siblings. An "item" was scored only once for each two-minute protocol, but the same subject might yield several items. While the results are only directional, they seem to show that opposite-sex siblings more often indulge in "impulsive" behavior.

Part of the interest of these data, if the direction is meaningful, is that it seems to imply that the subjects with opposite-sex siblings are more impulsive, though judging from the other categories, by no means out of control.

It makes sense to us to argue that this might be a mild form of the "impulsivity" that we have elsewhere shown to be associated with cross-sex identification (Sutton-Smith & Rosenberg, 1961a, 1961b)—an impulsivity which in its extreme forms is represented by the behavior-disrupting types of masculine protest also related to cross-sex identification in anthropological research (Harrington, 1969). Some further support for this association is offered by a pilot study in which we were assisted by one of our students, Stephen Goldman, who in 1968 observed the playground play of lower socioeconomic status Puerto Rican boys of same- and opposite-sex siblings. There were 24 five-year-old boys who were observed in alphabetical sequence in a playground period. They were observed according to the following code which was empirically derived during an initial period of observation.

I. Interaction with Adults

A. Attention-getting; controlling: 1. Angry crying; temper tantrum. 2. Asking for attention ("Look, Mira"). 3. Complaining about other children. 4. Being stubborn; refusing to respond.

B. Affection seeking; expression of dependency: 1. Touching (taking hand, tugging at pants leg or skirt). 2. Not eating. 3. Asking for things (toys, food, to have a story read). 4. Weeping, whimpering quietly, calling for mother.

II. Interaction with Peers

A. Aggressive or controlling: 1. Physical assertion (pushing, pulling, grabbing, assaulting). 2. Controlling, supervising without adult authority ("You put this away," taking ball from one child to give to another). 3. Nascent authoritarianism (taking things by force from smaller children—then appealing to adults for redress of grievances when other children apply force). 4. Verbal teasing; name calling.

 B. Affection seeking; expression of dependency: 1. Pleading, whining ("Just let me see"). 2. Crying as attempt to elicit nurturance from other children. 3. Asking for things or for help. 4. Following others.

 C. Friendly or cooperative activity: 1. Giving things to other children. 2. Putting things away after use without being told. 3. Spontaneously helping others to clean up or put things away, organizing group without being bossy ("Let's all get on line"). 4. Showing things.

The results are indicated in Table 3.14.

TABLE 3.14 Frequencies of Observed Behaviors

	M1M (3) M1F (3) (N = 6)	MM2 (2) FM2 (3) (N = 5)	Later Born Same-sex Sibs (N = 5)	Later Born Opposite-sex Sibs (N = 8)
To Adults:				
I. A. 1, 2, 3, 4	8	2	0	12
B. 1, 3	9	2	0	3
B. 2, 4	0	3	2	1
To Peers:				
II. A. 1, 2, 3, 4	20	4	6	12
B. 1, 3	3	2	2	1
B. 2, 4	0	2	1	1
Total	40	15	11	30

Again, the numbers are too small to provide anything but suggestive results, but once again those with opposite-sex siblings were more aggressive towards both adults and peers than those with same-sex siblings. Interestingly, the firstborn, as compared with second born, in the two-child families showed a similar trend. It will be recalled that in considering the data on cross-sex effects, it was suggested that part of the problem for firstborn males might be their forced dependency, and therefore their development of sex-role atypical traits.

 Once again, none of these data is sufficiently substantial to inspire great confidence. But like the data on creativity, these data on impulsivity seem to imply that there is some connection with opposite-sex role influences. At the extreme level, such opposite-sex role influences have been shown to be associated both with high-level creativity (Barron,

1963) and with delinquent masculine protest phenomena (Harrington, 1969). In the present work, what appears is perhaps a milder form of the masculine protest phenomena when boys with sisters report more masculine interests, fathers with daughters report more masculine interests, boys with sisters are impulsive in the classroom and, in the lower class Puerto Rican group studied, are more aggressive on the playground.

CONCLUSION

The present chapter has established that each sibling is affected by the sex of the other sibling, and that these effects are most obvious in the case of the younger siblings. It is the second born who most faithfully reproduce in their own behavior the responses which have been modeled for them (and perhaps demanded from them) by their older siblings. While the overwhelming evidence points to such direct influences of siblings upon each other, there is also a smaller body of evidence suggesting that subjects (particularly males) may counteract the opposite-sex role influence of their female siblings. There is also some evidence of the interacting effects of parental influence and sex role expectancies differentially affecting children of varying sibling statuses. While there is no direct evidence given here of the nature of these types of influence, the inference that they occur would hardly appear to be far-fetched. In the next chapter, we proceed to document that the older siblings do exercise more power over the younger siblings than vice versa, and can, therefore, be expected on theoretical grounds to exercise shaping and modeling influences.

sibling
power
effects

CHAPTER

4

The previous chapter considered examples of the way in which the sexes have a direct effect upon each other's interests and abilities. This chapter takes up more explicit examples of action and counteraction that are to be found in the power relationships between siblings. Like the previous chapter, it seeks an understanding of siblings in terms of their interactions, rather than in terms of their differential treatment by parents. From an Adlerian point of view, the materials of this chapter would perhaps be the most important because they deal with those naked power struggles that are said to have a direct impact upon all subsequent striving for superiority. From our own point of view, they are just one among the many classes of variables that play a part in creating differences between siblings. Still, they appear to be a particularly suitable illustration of direct sibling interactive effects.

POWER STUDIES WITH CHILDREN

The initial study of power by the present investigators (Sutton-Smith & Rosenberg, 1965b) began with children writing down their replies to two pilot questions about influence. These questions were chosen because a review of the psychology of power indicated that the concept of influence was the most central variable. The questions were (1) "How do you get your sibling to do what you want him (her) to do?" and (2) "How does your sibling get you to do what he (she) wants you to do?" Preliminary responses from four elementary fifth- and sixth-grade classes led to the development of a 40-item inventory reflecting the most frequently mentioned responses to these questions.

On each form, the child wrote his own name and the name and age of the sibling to whom he was making reference. "This is the way I get _____ to do what I want him (her) to do." Or: "This is the way _____ gets me to do what he (she) wants me to do." Firstborn children wrote in the names of their immediately younger siblings, and later borns wrote in the names of their immediately older siblings. For each item, the child was asked to circle the frequency with which that item was employed on a 1-point (never) to a 5-point scale (always).

Subjects in the study were 95 preadolescent fifth- and sixth-grade children at Kenwood School, Bowling Green, Ohio. Children in this school were of predominantly middle-class status. The types of items used in the inventories are illustrated in Tables 4.1 and 4.2 which give

TABLE 4.1 This Is the Way I Get My Sibling To Do What I Want Him To Do
(1–5 point scale—"never" to "always")

		MIM	MIF	MM2	FM2	FIF	FIM	FF2	MF2
1.	Beat up, belt, hit	3.00	1.80	3.06	2.50	2.44	2.33	2.67	2.22
2.	Promise	3.31	3.20	3.12	2.50	3.33	2.83	3.19	2.47
3.	Boss (say do it, shut up)	3.46	3.20	2.19	3.00	3.22	4.00	2.80	2.13
4.	Scratch, pinch, pull hair, bite	1.46	1.00	1.56	3.20	2.00	2.17	2.81	1.73
5.	Bribe, blackmail	3.23	2.60	3.44	3.60	3.56	2.33	2.52	2.60
6.	Ask, request	3.31	2.60	3.25	3.90	3.00	3.00	2.57	2.80
7.	Tickle	2.62	2.40	2.25	3.20	3.00	3.67	3.05	2.60
8.	Flattery	1.62	1.00	2.00	1.20	2.33	1.67	2.48	2.00
9.	Ask to do because older (younger, a boy or a girl)	2.23	2.40	1.94	1.70	3.22	2.83	1.71	2.47
10.	Wrestle, sit on, chase	3.31	3.60	2.25	2.70	2.89	2.67	3.10	1.86
11.	Bargain	3.54	1.80	3.19	2.00	2.67	2.83	2.95	2.33
12.	Ask parent for help	2.08	2.40	3.19	2.80	2.11	2.67	3.29	2.73
13.	Get angry (shout, scream, yell, get mad)	2.54	3.20	3.56	2.20	3.56	3.67	3.33	2.73
14.	Play trick	2.77	2.40	2.25	2.80	3.44	2.50	3.29	2.00

TABLE 4.1 *(Continued)*

		M1M	M1F	MM2	FM2	F1F	F1M	FF2	MF2
15.	Complain to parent	2.08	2.20	3.19	2.80	2.89	3.17	2.67	2.80
16.	Cry, pout, sulk	1.00	1.00	1.94	1.40	2.22	1.83	2.33	1.67
17.	Take turns	3.08	3.40	3.19	1.50	3.33	3.67	2.95	3.07
18.	Tell tales	1.92	1.20	2.13	1.50	1.32	1.83	2.95	1.47
19.	Attack things (hide toys, spoil bed)	2.38	2.20	2.31	2.50	2.67	2.00	2.38	1.80
20.	Explain, reason, persuade	3.46	2.80	3.12	2.90	2.89	3.17	3.14	2.73
21.	Ask other children for help	1.62	1.40	1.00	2.20	2.22	2.67	2.19	1.73
22.	Break things (toy, let air out of tires)	1.85	1.60	1.75	2.10	1.22	1.00	1.24	1.13
23.	Do something for the person	2.69	3.00	1.00	2.60	2.44	2.67	3.10	3.07
24.	Ask for sympathy	2.31	2.40	2.44	1.70	1.56	2.00	1.81	1.93
25.	Take things (ride bicycle, steal toys)	1.62	2.00	1.88	1.90	2.00	1.00	1.86	1.27
26.	Make feel guilty	1.92	1.60	2.62	2.00	2.00	1.00	2.38	1.73
27.	Bother (turn off radio, change TV channel)	2.77	2.60	3.50	3.30	3.11	2.83	2.81	2.60
28.	Pretend to be sick	1.62	1.60	1.63	1.80	1.56	1.33	1.48	1.47
29.	Use prayer	2.54	1.00	2.56	2.00	1.67	2.00	2.95	2.13
30.	Tease (name calling, pester, nag)	2.92	3.60	2.25	2.70	3.11	3.17	3.33	3.07
31.	Threaten to hurt	2.92	1.80	2.13	2.70	2.44	1.67	2.33	1.80
32.	Be stubborn, refuse to move	2.23	3.40	2.62	2.80	2.67	1.50	3.23	2.22
33.	Threaten to tell	2.31	3.00	3.50	2.80	3.44	2.83	3.00	2.40
34.	Make a wish about it	1.85	1.00	1.00	1.70	2.56	1.67	2.29	1.67
35.	Stop from using phone, bathroom, toys	1.62	1.80	1.19	1.90	2.33	1.50	2.71	2.07
36.	Spook them	2.77	2.60	1.69	2.70	2.44	1.17	2.81	1.20
37.	Exclude (can't play, can't go with—lock out of room)	2.62	1.80	1.94	2.40	2.67	2.50	2.52	2.07
38.	Give things (candy, money, toys)	3.23	1.40	2.19	2.00	2.00	2.33	2.57	2.27
39.	Give their choice (watch TV, play)	2.54	1.80	2.62	2.20	2.44	3.00	3.29	2.93
40.	Be nice, sweet talk	3.15	2.60	2.31	1.50	2.44	3.00	2.80	2.47

TABLE 4.2 This Is the Way My Sibling Gets Me To Do What He Wants Me To Do

		M1M	M1F	MM2	FM2	F1F	F1M	FF2	MF2
1.	Beat up, belt, hit	2.23	2.60	3.69	3.00	2.22	3.50	2.81	2.80
2.	Promise	3.31	2.60	2.38	1.80	2.67	2.17	2.95	2.00
3.	Boss (say do it, shut up)	1.85	3.40	3.69	3.40	1.22	2.33	3.52	3.13
4.	Scratch, pinch, pull hair, bite	2.08	3.40	1.75	2.80	3.11	2.67	2.52	1.67
5.	Bribe, blackmail	2.38	2.20	2.19	2.10	2.11	2.33	2.76	2.60
6.	Ask, request	3.15	2.40	2.44	3.10	2.44	2.50	2.76	2.40
7.	Tickle	2.38	2.40	2.19	2.30	2.89	1.67	2.86	2.27
8.	Flattery	1.85	1.80	1.50	1.20	2.22	1.33	3.00	1.47
9.	Ask to do because older (younger, a boy or a girl)	1.77	1.00	2.50	2.30	1.89	1.50	2.10	2.33
10.	Wrestle, sit on, chase	2.54	3.60	3.81	2.00	2.78	3.17	2.86	3.20
11.	Bargain	2.54	2.60	2.19	1.80	2.56	2.50	3.10	2.53
12.	Ask parent for help	3.00	3.60	1.94	2.80	3.44	3.50	2.24	1.47
13.	Get angry (shout, scream, yell, get mad)	3.08	3.60	2.50	3.80	3.33	3.67	3.00	2.22
14.	Play trick	2.54	1.80	2.69	1.90	2.78	2.33	2.57	1.86
15.	Complain to parent	2.46	4.50	2.13	3.20	2.33	3.50	2.81	1.27
16.	Cry, pout, sulk	2.31	2.40	1.69	1.70	3.67	2.33	2.43	1.27
17.	Take turns	2.85	2.80	2.56	1.60	2.11	2.33	2.48	2.13
18.	Tell tales	1.69	1.00	2.13	1.50	3.33	2.00	2.33	1.60
19.	Attack things (hide toys, spoil bed)	3.00	2.00	2.38	1.90	3.22	2.33	2.43	2.13
20.	Explain, reason, persuade	3.15	2.40	2.69	2.30	1.78	2.83	3.00	1.80
21.	Ask other children for help	3.23	3.00	1.94	2.00	2.44	2.83	1.86	1.53
22.	Break things (toys, let air out of tires)	1.85	1.80	1.19	1.80	1.78	1.67	1.33	1.67
23.	Do something for the person	2.38	1.80	2.25	1.60	2.33	3.60	2.80	2.33
24.	Ask for sympathy	1.85	2.20	1.38	1.80	2.44	1.83	1.91	1.67
25.	Take things (ride bicycle, steal toys)	2.15	2.20	2.31	1.50	2.22	2.50	2.24	1.86
26.	Make feel guilty	1.85	1.40	1.19	1.40	3.22	2.17	2.48	2.07
27.	Bother (turn off radio, change TV channel)	3.08	3.60	3.50	3.10	2.67	3.50	3.10	2.80
28.	Pretend to be sick	1.62	2.00	1.44	1.50	2.11	1.67	1.52	1.47
29.	Use prayer	1.85	1.80	1.19	1.80	1.44	1.67	2.71	1.80

TABLE 4.2 *(Continued)*

		M1M	M1F	MM2	FM2	F1F	F1M	FF2	MF2
30.	Tease (name calling, pester, nag)	3.38	2.60	3.50	3.00	3.67	3.33	3.05	3.27
31.	Threaten to hurt	1.46	2.60	2.31	2.40	2.33	1.50	2.10	2.00
32.	Be stubborn, refuse to move	2.62	2.00	2.25	2.00	3.44	2.83	3.05	2.00
33.	Threaten to tell	3.38	3.20	3.06	3.20	2.67	3.33	3.43	1.86
34.	Make a wish about it	1.62	1.00	1.56	1.70	2.56	2.00	2.10	1.07
35.	Stop from using phone, bathroom, toys	1.31	1.40	1.75	2.10	2.33	1.17	2.67	2.27
36.	Spook them	1.85	1.80	1.88	1.40	1.67	1.17	2.23	1.47
37.	Exclude (can't play, can't go with, lock out of room)	2.23	2.00	3.06	2.60	2.44	2.67	2.57	1.93
38.	Give things (candy, money, toys)	2.08	3.20	1.88	2.00	2.11	2.17	2.43	2.47
39.	Give their choice (watch TV, play)	2.23	2.20	2.25	2.40	1.89	2.67	2.76	2.80
40.	Be nice, sweet talk	2.69	2.80	2.31	1.50	1.78	2.33	2.71	2.27

the average score for each sibling position, both on influencing and being influenced. Because a novel scale of this sort is susceptible to as yet unanalyzed response sets, it was decided to conduct the analysis on those items on which the siblings of complementary ordinal relationships indicated consensus. For example, if the older sibling rated himself as more bossy than the younger, and rated the younger one as less bossy than himself, and then the younger also stated that he was less bossy than the older sibling, and the older was more bossy than he, there was a consensus across the four responses to the one item "bossy." The statistical technique used to arrive at such a consensus between siblings was a factorial analysis for unequal cell frequencies with repeated measures using an unweighted means solution (Sutton-Smith & Rosenberg, 1968).

The data stated in Tables 4.1 and 4.2 indicate items on which significant differences were obtained in the siblings' mutual perception of each other. Each dyad is considered separately: older brother and younger brother (M1M and MM2); older sister and younger sister (F1F and FF2); older brother and younger sister (M1F and MF2); and older sister and younger brother (F1M and FM2). Levels of significance as low as $p < .25$ are occasionally indicated because they parallel differences of the same content but higher significance level for some of the other dyads. (See Table 4.3.)

TABLE 4.3 Reciprocated Interactions in Sibling Dyads: Children $(N = 95)$ (Analysis of Variance, Position by Form)

M1M	MM2
Beat up, belt, hit (.01)	Get angry, shout, yell (.05)
Boss (.001)	Cry, pout, sulk (.01)
Threaten to hurt (.01)	Ask other children (.01)
Spook (.05)	Threaten to tell (.05)
(Exclude) (.10)	(Complain to parent) (.10)

F1F	FF2
Boss (.001)	Scratch, pinch (.05)
Ask to do because girl (.05)	Ask parents for help (.01)
Explain reason (.05)	Cry, pout, sulk (.05)
(Take turns) (.10)	Tell tales (.05)

F1M	FM2
Boss (.05)	Bribery, blackmail (.05)
Tickle (.05)	Wrestle and chase (.05)
Ask (.01)	Break things (.05)
Get angry (.05)	Take things (.05)
Take turns (.05)	Make feel guilty (.05)

M1F	MF2
(Tease) (.10)	Scratch, pinch (.001)
Stubborn (.05)	Ask parent for help (.05)
(Boss) (.25)	Complain to parent (.001)
(Wrestle, chase) (.25)	Cry, pout, sulk (.01)
(Play tricks) (.25)	Ask other children (.05)
(Spook) (.25)	Use prayer (.05)
	(Flattery) (.10)

Source: Sutton-Smith, B. & Rosenberg, B. G. Sibling consensus on power tactics. *Journal of Genetic Psychology*, 1968, **112**, 63–72.

The greatest consensual agreement in the results is that firstborn are perceived both by firstborn and second born as more bossy: M1M $(p<.001)$, F1F $(p<.001)$, F1M $(p<.05)$; M1F $(p<.25)$. Similarly the non-firstborn tend to show what is presumably a low-power procedure of appealing to others outside the sibling dyad for help: crying, pouting, sulking, or threatening to tell tales or using prayer. Thus, MM2 ask other children $(p<.01)$, complain to the parent $(p<.10)$; FF2 ask parents for help $(p<.01)$, tell tales $(p<.05)$; MF2 ask parents for help $(p<.05)$, complain to parents $(p<.001)$, ask other children $(p<.05)$. Crying and pouting are also common to MM2 $(p<.01)$, FF2 $(p<.05)$, and MF2 $(p<.01)$. Only the boy with an older sister does not seek help outside the dyad, nor does he sulk or pout. Apparently, the younger boy with an older sister does not show the same degree of powerlessness as the other non-firstborn categories, a finding which replicates a difference for these boys in an earlier study (Sutton-Smith, 1966). His use of bribery, blackmail ($p<.05$), breaking things ($p<.05$), taking things ($p<.05$), making the F1M (his reciprocal in the dyad) feel guilty ($p<.05$) are unique among the non-firstborn.

Physical power tactics appear to vary with sex. Thus, beating up, belting and hitting (M1M, $p<.01$), and wrestling and chasing (FM2, $p<.05$), are ascribed to boys. But scratching and pinching (FF2, $p<.05$; MF2, $p<.01$) and tickling (F1M, $p<.05$) are ascribed to girls. The direct physical power of M1M and FM2 on their sibling opposites apparently has a strong effect, as their partners are the only ones who score higher on getting angry, shouting, and yelling (MM2, $p<.05$; F1M, $p<.05$).

The polite and perhaps strategic techniques of explaining, asking, and taking turns are attributed only to firstborn girls (F1F, $p<.05$; F1M, $p<.01$).

The firstborn boy with the younger sister is not as clearly characterized as the others, but his attributes (directional, but nonsignificant) are similar to those of the other firstborn boy (M1M) (boss, wrestle, play tricks, and spook).

The quality of some of these responses is gained from an interview study in 1966 by these investigators with the assistance of Yaël Orbach involving children who gave detailed responses when asked how they bossed, blackmailed, or sulked, and so on. We list below examples of their responses. All names have been changed.

Reason

(How does your brother get you to do what he wants you to do?) Well, he wanted me to get his binoculars. And I said, "Well, what do you want them for?" And he said, "Well, I want to look at this neat, little bird up there." So I brought him his binoculars. And he showed me the bird. And then he went off. (MF2)

Well, we were both going to practice at the same time—her on the violin and me on the piano. And we had to sit down and say whose was more important. And we finally decided that mine was a little more important because I have an hour to do. (F1F)

Obligation

(How does John get you to do what he wants you to do?) Well, he says, "Well, I did that for you. Now you do this for me." So I'm just stuck. (MF2)

Well, once in a while she likes to go to movies or something and I'll give her money. And I'll say, "You've gotta help me or something someday!" And the time comes when I want some help and I want her to help me do the dishes or help me with the paper route or something on Saturday morning or something like that. And I'll talk to her about it. I say, "Well, come on, Jill. I let you have some money. And I've helped you a lot and the least you could do is help me when I need help!" And she usually helps me. Once in a while she says, "I'm busy," or "I have to go," or something like that but usually she helps. (M1F)

Request

(How do you get Donald to do what you want him to do?) Oh, a little blackmail and bribe. Most of the time I can just coax him. Well, this fall we were playing—I wanted to play football. And he, at the time, was not very interested in playing things. But I coaxed him into playing with me, and we had a lot of fun. (M1M)

Bribe

(How do you get Kay to do what you want her to do?) If I want her to go uptown with me, I tell her that I'll give her the change. If I want her to do something for me, I'll tell her that I'll play house with her. (F1F)

(How do you get Kate to do what you want her to do?) Well, if she won't do it, I usually say, "You can hold Cuddles!" That's my cat. (M1F)

Boss

(How does Mary get you to do what she wants you to do?) By telling me. She'll just tell me and I'll do it. Well, I have to leave for school at 8:00 and a lot of times I've been getting up at a quarter to eight. And she'll tell me I gotta make my bed. And I don't have time to do it. What does she do then? I forget. Well, she'll keep me after school. I mean when I get home from school, she'll keep me in the house 'til I do clean my room and stuff. (FF2)

(Did you ever try to get Pat to do what you wanted by bossing and shouting and getting angry with her?) Yes. Once I wanted to use one of her pens for a report I had and she wouldn't let me use it. And I said, "You better give it to me." Then she said, "All right." She's just trying, you know, not to give it to me to see how I boss her around. (FF2)

Attack person

(Did Lois ever try to get you to do what she wanted by sulking or pleading or whining?) Well, she doesn't usually plead. She sulks a lot. I said that I wouldn't call Laura because I don't know her number and so she began hitting me and hitting me and hitting me until I had a great big black-and-blue spot right on my leg. No, I guess it wasn't on my leg. I guess it was on my arm. And I can't remember where it was. Anyways, I finally gave in. (F1F)

(Did Pam ever try to get you to do what she wanted by attacking you or holding you?) Yes. Well, yes sometimes she tries to tickle me, and I'm terribly ticklish. And so she'd try to tickle me and like sits on top of me and starts tickling me. And finally I have to do what she wants. (F1F)

Attack property

(What is the worst thing that Molly ever did to you?) Took all of my marbles. She took all of my marbles. (MM2F)

(What is the worst thing that Nancy ever did to you?) I don't know. Maybe something bad? Well, one time she got in my room and I was mad at her so she wrecked my bed after I had just made it real nice. And she messed up my shelf and everything. *(What is the worst thing that you ever did to her?)* I took her doll. She starting crying. (F1F)

Shout

(Did you ever try to get her to do what you wanted by bossing and shouting and getting angry with her?) Sometimes. I told her to get out of my room. And I kept shouting at her and she wouldn't go. And I started hitting her, and she still wouldn't go. So I just picked her up and threw her out. (M1F)

(Did Lois every try to get you to do what she wanted by bossing and shouting and getting angry with you?) Yes. Well, one day she was in a real crazy mood. She had just gotten home from school and she couldn't call up her friend Ann. And, oh, all this junk. And she'd keep saying, "I want to!" and, "I'm gonna do it! And I'm gonna do this and I'm gonna do that!" And, then finally I got so mad at her, I just yelled at her and I said, "Be quiet!" And she said, "Well, if you're gonna be that way, clean up my room. It's messy!" (F1F)

Withdrawal

(When things go wrong with Joan—she gets mad at you or cries, or gets hurt, or won't speak—what do you do?) Leave her alone. Well, I threw rags all over her room and she got real mad and she wouldn't tell or talk to me. So I said, "I'll clean them up!" And I walked into her room and she scratched me across the arm. And it started bleeding. So I'm not going near her when she's mad. I'll leave her alone. (F1F)

(What do you do when your parents get mad?) I'm not around. I always go up to my room and lock me in. Just sorta awful. Nothing, I just sit around until they stop yelling at me. (FM2)

Harassment

(Did you ever try to get Jane to do what you wanted by teasing, pestering, harassing?) Yes. I wanted to stay up and clean the bedroom and she didn't want to, so I told her she'd turn into a hog and would turn brown and that she'd get in trouble. (F1F)

(Did Bob ever try to get you to do what he wanted by teasing, pestering, harassing?) Yes. He wanted to play baseball and I didn't want to so he started teasing me, like saying, "What's the matter? Can't you pitch? Can't you catch?" (MM2)

(Did you ever try to get her to do what you wanted by teasing, pestering, harassing?) Yes. Just an example, but this summer this boy came over to play and everything, with his family. And they went out in the woods to

hunt for something. And she wouldn't help me with whatever I had to do. I said, "I'll tell Daddy that she's out there with that boy and everything, kissing!" (FF2)

Plead

(*Did Nelson ever try to get you to do what he wanted by sulking or pleading?*) Once in a while he has. Well, last night we were playing cards and he said, "I don't want to play this anymore. I want to read with you!" He didn't want to play cards; he wanted to read. And so I said, "Well, maybe, but I don't think so!" And then he started saying, "Oh, come on, pretty please!" And finally I had to give in. (F1M)

(*Did Gwen ever try to get you to do what she wanted by sulking or pleading or whining?*) Not by sulking, but she might plead. I mean like maybe she might say, "Oh, please, come on, Ruth. I need it"—or something like that. (*Could you think of an example?*) Well, maybe she wanted to put on her eye shadow or something or eye make-up. And she didn't want to get it because maybe she was still washing her face or something, you know. "Oh, come on please, Ruth,"—you know. "I'll do it for you, you know, when you get older if you please just get it for me just now." (FF2)

Trickery

(*Did Shirley ever try to get you to do what she wanted by teasing, pestering, harassing?*) Yes. Well, if she wants me to help her carry, oh, a box or something down to the basement, she says, "Well, there's something in my room that I want you to see!" And I finally go into her room and I see it. And it's the box. And she says, "Okay, you saw the box; take it downstairs!" (FM2)

(*Did Mark ever try to get you to do what he wanted by making you feel you owed it to him, you should do it because of the things he had done for you?*) Yes. Well, one time I owed him some money and I hadn't quite paid it back yet, and so he said, "Connie, if you make my bed, you could forget about the money!" And so I made his bed. And he says, "Connie, you still owe me some money!" That makes you mad, too. (F1M)

Tell parent

(*How does Mona get you to do what she wants you to do?*) Well, sometimes she would say, "If you don't do that, I'll go tell Mom!" Well, one time she wanted me to get her this book. And I said, "Well, why don't you get it yourself: you have two legs!" And she said, "Okay, I'll just go tell Mom that you haven't been nice to me!" (MFF3)

(*How does Miriam get you to do what she wants you to do?*) She usually says that she'll tell Mom or Dad. Like, well, one night I stayed over with one of my girl friends—overnight—and I used her Barbie case. I dumped all her doll clothes out in a box and when I got back, she wanted me to put them back. And I didn't want to, so she started to tell Mom. So I had to do it. (F1F)

The interview was given to 68 boys (40 firstborn and 28 later born) and 72 girls (34 firstborn and 38 later born) in the third, fourth, and fifth grades of Kenwood School, Bowling Green, Ohio. In many ways it repeated the data already gathered in an earlier year using the inventory mentioned above. However, new questions were added, and by taping the children's responses we gained an enriched understanding of the results reported from the inventories. In the material below, gross contrasts are made between firstborn and later born in terms of various content analyses of the children's answers. Although records were kept of the specific ordinal positions, and although these and family size were balanced in such comparisons, there were insufficient numbers in all categories to allow for more specific contrasts. Given the ad hoc nature of the categories and the failure of all respondents to fit each analytic system, the data are presented below without statistical tests. They are of suggestive value only, though the findings often parallel those already presented.

The following questions were a part of the interview:

1. *If you could be one of your siblings would you want to change?*

The majority did not want to change. Only one or two in each ordinal category indicated any interest, except for later-born girls with older sisters, among whom 8 of 23 said they wanted to change. It will be remembered that these were the female siblings who seemed to fare less well in the cross-sex data presented earlier.

2. *Which member of your siblings has the most fun?*

As the figures in Table 4.4 indicate, males tend to favor later-born positions, and females, to favor firstborn positions, though this difference is more pronounced among the later born. The small number of children not included in the table said that they had equal fun with their sibling. As mentioned in the previous chapter, the males' view that the later born have more fun may be associated with the sex role typical independence of that position; also, the females' view that the firstborn

TABLE 4.4 Who Has the Most Fun?

| | Answers of: | | | |
	1B Male (N = 40)	LB Male (N = 28)	1B Female (N = 34)	LB Female (N = 38)
Older have more fun	11	9	16	26
Younger have more fun	28	18	14	10

NOTE: 1B = firstborn
LB = later born

have more fun may be associated with the sex role typical parental dependence of that position.

3. *How do you have fun with your sibling?*

At least half the responses for all ordinal categories involved games. Older brothers and younger sisters tended to mention going places with the sibling. Females more than males mentioned such items as helping, teaching, reading, cleaning, doing homework, going to the library, sewing, and dancing. Males gave a more exclusive emphasis to recreational pursuits.

4. *When things go wrong with your sibling—he gets mad at you or cries or gets hurt or won't speak—what do you do?*

There were not many differences here, except perhaps a slightly stronger tendency for male firstborn to say that they would make up. Subjects could respond in several categories (see Table 4.5).

TABLE 4.5 Handling Siblings

	1B Males	LB Males	1B Females	LB Females
Ignore	15	13	19	26
Get mad	7	5	10	13
Make up	16	9	9	7

When asked the reverse question of what the sibling did when he or she got mad, there were few differences except that the firstborn claimed that the later born more often told parents. There was a tendency also for the firstborn to make up or apologize if the younger one was a girl rather than a boy.

5. *What is the worst thing your sibling ever did to you?*

Again, there were only slight differences, with firstborn declaring that the later born do more property damage, and the later born declaring that the firstborn interfere with them more. The items offered by the children are of interest: (a) *physical attack*—Hit me, chased me, locked me up, tickled me, yelled, threatened; (b) *property damage*—Broke something, took something, messed my room, messed my things, hid something of mine; (c) *interference and disagreement*—Changed TV channel, wouldn't let me sleep in his bed, didn't do it and I had to, tacked a lot of work on me, wouldn't let me go out, bothered me, quit in the middle of a game, made me mad, had an argument, didn't let me play, didn't let me use something; (d) *embarrassment*—Told on me, embarrassed me, called me names, made fun of me, broke a secret, broke a promise, cheated, got me in trouble with parents. The items and categories may be open to dispute, but the flavor of sibling relationships is indisputable.

6. *What does your sibling do when you get mad?*

Responses were divided into four categories: (a) *apologizing* (tries to make up, explains, apologizes); (b) *ignoring* (leaves me alone, doesn't play, doesn't talk, goes to play with someone else); (c) *offensive reactions* (hits, gets mad, tries to make me madder, calls names); (d) *defensive reactions* (runs away, tells Mom, cries). Subjects often responded in several categories (see Table 4.6).

TABLE 4.6 Response to Being Mad

	Firstborn Report on Later Born		Laterborn Report on Firstborn	
	1BM	LBM	1BF	LBF
Apologize	8	11	6	7
Ignore	12	16	19	30
Offensive	24	16	23	32
Defensive	20	4	14	6

Firstborn males consider the later born both more offensive and defensive. Firstborn girls see the later born as more defensive, and the later born see the firstborn as ignoring them more and being more offensive. Within subcategories, later-born males are usually described as more offensive, and later-born females as defensive. A later born of the same sex is as offensive, and of the opposite sex, defensive. These are sex and ordinal differences that might be expected from the study of power differences discussed earlier.

7. *How do you get your sibling to do what you want him to do?*

The results are not dissimilar from those derived from the inventories. Older siblings attack, use status and bribe more, whereas later born plead more. In addition, there are sex differences. Females contend that their sisters use more bribes, in contrast to their brothers, who use more requests (see Table 4.7).

TABLE 4.7 Influence Techniques

	1BM	LBM	1BF	LBF
Use physical attack	7	1	—	—
Get mad, yell	—	2	2	2
Appeal to status	5	1	2	—
Bribe, blackmail	32	18	29	34
Ask, request	26	21	16	21
Plead, cry	3	21	8	16

A series of questions on whether a sibling used a particular tactic (say, bossing) with the other sibling yielded the following results:

Bossing—Firstborn males and females reported being more bossy with a next-younger female than a next-younger male. Later-born males and females reported being more bossy toward a next-older sibling of the same sex than toward a next-older sibling of the opposite sex.

Reasoning—Later-born females reported more reasoning from older sisters than from older brothers. They also claimed they used more reasoning with older sisters than with older brothers.

Attack—Later-born females reported more attacks from older males than from older females.

Sulking—Firstborn males contended later-born males sulked more than later-born females. But later-born females report more sulking by older females than older males. Sulking appears to be a technique most often used with the same-sex sibling.

Obligation—This is a technique reported by older and younger sisters for use with each other, rather than by males or opposite-sex siblings.

Teasing—Teasing is used by same-sex siblings with each other.

From these tentative results it appears that there are some tactics that are most often used by males (attack, offense); some most often used by females (reasoning, defense, making the sibling feel obligated); some more typical of firstborn (interfere, ignore, be offensive, attack, bribe, use status, and boss); and others more typical of second born (attack property, plead, reason). Again, some seem more often to be used with the same sex (offense, boss, sulk, and tease), while others are more often used with the opposite sex (make up, defense, and boss). As in the inventory study, males use more physical techniques as compared with the more symbolic techniques of females. Firstborn use more powerful techniques than later born, and same-sex siblings appear to indulge in more powerful interactions than do opposite-sex siblings. The relationships of the latter may be moderated by codes concerning heterosexual relationships.

Several questions about *parents* yielded fairly conventional expectations. Asked about parents getting mad, more firstborn reported acquiescing and apologizing and more later born reported reacting with anger. Firstborn girls contended that later-born girls more often cried, while the later-born girls said the firstborn went off and isolated themselves. Asked which parent they had the most fun with, girls more often chose mothers, and boys more often chose fathers, but girls were more inclined than boys not to choose and to claim equal fun with both parents. Among males, firstborn more often chose mothers. Females with sisters felt relatively closer to mothers, while females with brothers felt relatively closer to fathers.

Questions about power tactics used with parents indicated that pleading was more often used by females, and used by both sexes with the mother rather than the father. The males more often saw the mother as getting mad, yelling and screaming at them. Both sexes saw the parents as more bossy with siblings of the same sex than with siblings of the opposite sex.

These interview materials can be compared with Koch's interview study with 6-year-olds who were asked how much quarreling they did with their siblings (1960). The rank order and the percentage with which the siblings mentioned some quarreling was as follows: FM2 37.7 percent; M1M 35.3 percent; MM2 33.3 percent; FF2 29.7 percent; F1M 29.7 percent; MF2 29.0 percent; M1F 20.3 percent; F1F 13.3 percent (1960, p. 26). It is noticeable that the second boy with the highest power tactics (FM2) also reported more quarreling. He seems to be the second born who has the greatest chance of overthrowing his elder sister's power. Koch has discussed this older sister (F1M) as being challenged by the younger brother's sex identity. It is clear he challenges her quite directly also. The all-boy dyad reports the greatest dyadic amount of quarreling. The M1F and F1F who report the lowest amount of quarreling also yielded fewer significant power relations in the present preadolescent study. The percentage of times in which the subject admitted that the sibling was usually victorious seems to be fairly closely related to the power reports also. Thus: FF2 25.3 percent; MM2 15.7 percent; M1F 13.7 percent; FM2 13.3 percent; M1M 11.3 percent; MF2 11.0 percent; F1M 9 percent; F1F 7 percent. Second borns confess defeat more readily, just as they admit quarreling more.

With respect to the introductory theoretical material, in which we stressed the importance of considering sibling as well as parent influences, it can be said that some of the differences agreed upon by siblings in the present study appear to derive from parental sources and that some appear to derive from what might be termed the character of the sibling interaction and the inevitable differences in power between siblings, given their differences in age and size. Taking first those inventory-derived characteristics of the firstborn that might be modeled after their parents: the older girls' use of explaining, giving reasons, and asking may perhaps be modeled after that of their mothers, with whom they have closer relations than do the second born. Again, "excluding" and "taking turns" might perhaps be regarded as "affiliation" strategies (see Chapter 5), assuming that those (firstborns) who are more concerned and anxious about affiliation learn to develop such devices for its management.

Some of the other findings are perhaps best interpreted in terms of the intrinsic characteristics of high- and low-power relationships, generally, rather than in terms of the relationships between the siblings and their parents. No matter whether the group is an animal or a human one, those who are larger in size and ability usually exercise dominance (bossiness) in order to ensure themselves of greater access to the available rewards. Dominance exercised by the more powerful, and anger and resentment (shouting, yelling) exercised by the less powerful may be regarded as universals across social systems. In the human family system, of course, one of the available rewards is the intangible affection of the parent, so that the exercise of power by the firstborn may be greatly increased by his jealousy of the non-firstborn as a baby. A long literature on sibling rivalry attests that sometimes such jealousy can instigate increased aggression and dominance by the firstborn (Levy, 1937). The point being made here, however, is that although this peculiarly human jealousy increases the dominance of the older sibling, it is not the origin of such dominance. The accident of birth creates size and ability differences, and these no less than size differences in lower species lead to the institution of a pecking order. We may assume the less powerful, younger siblings will fight back with all the powers at their command. The younger brother's relative strength, when he has only an older sister to contend with, may be taken as an illustration of this view. Usually, however, younger siblings seem to have to be content with the exercise of greater power outside their own sibling groups (Veroff, 1957; Krout, 1939; Singer, 1964; Sutton-Smith, 1966b). Perhaps, as Henry (1957) has said, power needs (if not relations) are symmetrical; that is, those seriously overpowered have needs to repeat the treatment to some other persons less powerful than themselves. Much of the literature on the authoritarian personality seems to be based on such a premise.

Again, looking at the sibling group as a miniature social system, the appeal by the non-firstborn for the support of their parents (threaten to tell, complain to parent, ask help from parent, ask other children) parallels the tendency in politics and in small group research for the weak to ally themselves with a stronger third force. This strategy appears to have much generality as a group relational characteristic (though a hardly advisable one in politics according to Machiavelli). In the family, this characteristic is, in part, supported by the parents' being more indulgent with the younger children and thus implicitly encouraging their appeal for help (Lasko, 1954; Bossard & Boll, 1956b).

Yet another way of looking at the sibling social system is to consider it, with its dominance hierarchy, as a system in a state of balance. As long as older and younger stay with their respective powers, there is a minimum of fighting. According to Scott (1963), this is one of the functional virtues of pecking orders at the animal level. They minimize conflict. Yet, if we pay attention to the harassing by the younger ones, and to their attempts to appeal for assistance, we can add that they seem to be attempting some of the time to destroy this system of balance with their siblings. Thus if they can prove more suffering (by complaining or crying, for instance), they can usually get the parents as a third party to interfere and at least temporarily upset the system. And they can often do this by first teasing and pestering the older ones until, losing their self-control, the older ones make use of excessive power, so that appeal by the younger ones to their parents is then legitimized. If this is the correct description of the state of affairs, we may perhaps postulate that if dominance hierarchies reduce overt hostilities, at the same time they also induce, in the low-powered members, countercoalitional activity, aimed at upsetting the established order.

Another set of data which is relevant to the understanding of these power differences is available from the Berkeley Guidance Study the data from which was made available to these investigators for reanalysis (MacFarlane, 1938). This study included ratings made on children between the ages of 6 and 16 years on the dimension of *domineering-submission*. The ratings were made by a case worker on the basis of a variety of types of information including teachers' observations, parental reports, and interviews with the children. The sample included nine boys with older brothers, eight boys with older sisters, five boys with younger brothers, and eleven boys with younger sisters ($N=33$). There were ten girls with older brothers, nine with older sisters, five with younger brothers, and five with younger sisters ($N=29$). The small size of the sample makes very tentative any generalizations that can be drawn from these data. The clearest difference is between the ratings of the girls with younger brothers (F1M), who received the lowest ratings on domineering throughout, and their younger brothers (FM2), who received the highest ratings throughout. This finding is consistent with other data presented above. Whereas the F1M and FM2 thus pursue complementary paths—one being submissive, the other dominant— the other opposite-sex pair, M1F and MF2, shows a very similar series of ratings to each other over these age levels, with the MF2 getting slightly higher ratings throughout. She parallels the behavior of her older brother, whereas the younger brother with an older sister counteracts it —a further example, it seems to us, of the different ways in which the sexes affect each other already noted in the previous chapter. The scores for the same-sex siblings are not as clearly distinguished across the ages 6 to 16.

POWER STUDIES WITH COLLEGE STUDENTS

The initial study with college students was similar in character to those carried out with the children (Sutton-Smith & Rosenberg, 1965b). The inquiry began with an open-ended questionnaire about influence, but this time it included not only influence upon and by siblings but also influence upon and by mothers, and upon and by fathers. Because the students' replies were in paragraph form, a content analysis of these replies by sibling position was possible. In making the analysis, the assumption was made that the types of power could be scaled from high to low.

At the high end were those techniques reported significantly more often by males than by females, and more often by firstborn than by second born, and more often by parents than by children. At the high end of the scale were such techniques as command, physical restraint, making the other feel guilty, making the other feel obligated, glorifying the job to be done, depriving of privileges, and reprimanding.

At the low end were items reported significantly more often by females, second born, and children: sulking, praying, wishful thinking, claiming illness, pleading and whining, appealing for sympathy, appealing to parents, threatening to tell, teasing, and harassing.

In the middle was a no man's land of nondifferentiating items where those of high or low status could resort to reason, explanation, flattery, bargaining, bribery, trickery, blackmail, anger, and shouting.

In these terms, children in only-child families were found to describe their parents as more similar to each other in the array of power techniques that they used and as using more high-power techniques than did parents in two-child families, as described by these children. Fauls and Smith (1956) present similar data. In two-child families, the mother and father were more differentiated, with the father described as using more high-power techniques and the mother, more low-power techniques. We might conclude from this that the children are relatively more powerful in the two-child family.

In the two-child family, the females reported using more differentiated high- and low-power techniques with mothers and fathers than did the males. Males reported using trickery and appeal to the fact that they were boys, with both mothers and fathers; whereas girls reported pleading, whining, and appealing for sympathy with the mother, but reasoning, persuading, and giving rewards to the father.

Both sexes were similar in perceiving the opposite-sex parents as using an appeal to their sex (do it because you are a girl—a boy), and the same-sex parents as using a greater variety of pressure tactics. Boys saw their fathers in particular as using physical restraint and blackmail, and girls saw their mothers as reprimanding them, glorifying the job to be done, making them feel guilty about it, making them feel obligated to do it, and appealing for their sympathy. Mothers, like daughters, also were seen as using a wider range of power techniques.

As the literature would lead one to suspect, firstborn as compared with second born reported many more differentiated types of influence upon and influence by parents. The second born, on the contrary, claimed to suffer a more homogeneous administration of high-power techniques. In the face of the superior power of both parents and first-born, the second borns' technique most often used was that of withdrawal, a finding reminiscent of Schachter's work on affiliation (1959).

From the preliminary content analysis, a list of inventory items was once again compiled, as had been done in the case of the children. This 43-item inventory—in six forms (influencing sibling, being influenced by sibling, influencing parent, being influenced by each parent)—was then submitted to 15 subjects in each of the 8 two-child family positions. Subjects were 19 years of age and of similar socioeconomic status. As before, the responses of siblings in the same family dyads (M1M and MM2) were subjected to analysis of variance to find which statements of the subject about himself and his sibling were consistent with the statements of the sibling about himself and the subject. Consensus was achieved on about 25 percent of the items in the inventory in each dyad, though the siblings of the same sex showed a higher consistency than those of the opposite sex.

Consensually validated findings of this sort at both the college and the child levels were as follows: All firstborn perceived themselves and were perceived by second born as exercising higher power—that is, they commanded, reprimanded, scolded, and bossed. Reciprocally, the second born pleaded, whined, sulked, and appealed for help and sympathy from the firstborn and others. Similarly, the firstborn used more physical restraint and physical attack, though there had been more of this at the child level, where it had involved beating up, wrestling, and hurting. The firstborn also gave more rewards and deprived the second born of more privileges. The second born responded by getting angry, being stubborn, and by harassing, pestering, and bothering the firstborn.

Intrigued by the degree of consensuality, we hypothesized that if firstborn and non-firstborn were put into an experimental role-playing setting and were required to play "bossy" and "bothersome" roles, respectively, the firstborn would do better at the bossy role (a firstborn technique), and the second born, at the bothersome role (a second-born technique). Ten pairs of firstborns and second borns (girls) played against each other (F1F and FF2). They were observed by ten pairs of similar firstborn and second born through a one-way screen, who, without knowing their ordinal position, subsequently rated their effectiveness in playing bossy and bothersome roles. The roles of the players were altered so that firstborn played the bossy roles five times and the bothersome role five times. The same was true of second born. In each case for a pair, one would play one role, while the reciprocal (the opposite-numbered role player) took the opposite role. The Ss thus included ten F1F's and their counterpart FF2's as role players, and ten F1F's and ten FF2's as judges.

The instructions to role players were as follows: You are going to be asked to play a role with another person who will play the role of your sister. You are trying to decide how you will spend your vacation together. You both disagree on how this will be done. You are going to try to get your sister to do what *you* want her to do. You are to disagree on any suggestion your sister makes. *Instructions to Player A:* We want you to play your part in a bossy manner. You are to be authoritative, commanding, and reprimanding. *Instructions to Player B:* We want you to play your role in a bothersome manner. You are to harass, nag, and pester the other person. *Instructions to judges:* You will witness two persons playing a role. They will play the role of two sisters trying to decide how to spend their vacation together. They will both disagree on what they will do, and each will try to get the other to do what she wants to do. The one on the right will play a bossy role; the one on the left will play a nagging role. You are to judge how effective they are in playing their roles. We expect they will differ in their ability to play these roles. *Instructions to judges regarding ratings:* Here are a set of rating scales for you to judge the two role players. First, on a 1- to 9-point scale, how effectively did the one on the right play a bossy role? How effective was the one on the left in playing the nagging role? In viewing the role playing, with whom did you identify? Who had the best dialogue? Whose gestures and expressions were best?

The results indicated significant agreement that the non-firstborn were perceived as superior role players, a point to which we will return in Chapter 8. The results showed also that firstborn raters identified with the bossy role (whether played by firstborn or second born), and the non-firstborn identified with the bothersome role (whether played by firstborn or second born). In sum, while there was consensual validation across the F1F's and FF2's as to the superiority of the FF2's as role players, each nevertheless showed a preference for the role previously predicted to be her own, the bossy or bothersome role, respectively (Sutton-Smith & Rosenberg, 1966a).

This finding was of considerable importance if it meant what it seemed to imply—that adult preference for others was to a considerable extent consistent with enduring sibling-association influences throughout the growth years. It might be important to know that a considerable portion of the population has a built-in disposition to feel most identified with the bossy one, and another segment has a built-in disposition to identify with the bothersome sibling! Or, if you will, it reflects a disposition to identify with messiahs on one hand and with agitators on the other—which is the way Harris (1964) might state it. Unfortunately for science and our enthusiasm, the results were not replicated in a similar and subsequent experiment. Nevertheless, the general durability of these findings throughout the age levels indicated that a further pursuit of the character of power relationships derived from these inventories would be worthwhile.

In searching for superordinate concepts to take us beyond the material we had collected on particular tactics, we had various alternatives. We could have made a special grouping of items in terms of some of the classifications in the literature, such as Singer's "Machiavellianism" (1964); French and Raven's notions of "coercive power," "legitimate power," "referent," "reward," and "expert" power (1959); or, "strategic," "potent," and "fortunist power" (Sutton-Smith & Roberts, 1964). But it seemed wiser to proceed initially in an inductive way with our materials. Table 4.8 presents the results of separate factor analyses of male and female responses to our inventory, regardless of ordinal position, and using the principal axes technique.* The respondents were 120 19-year-old college sophomores (60 male, 60 female), members of the 8 types of two-child family (15 from each group, M1M, MM2, etc.) and of predominantly middle socioeconomic status. What we have in this study, in effect, is a factor-analytic study of sex differences and similarities in the use of power tactics with siblings. The factors and their items loadings are indicated in Table 4.8. What the table shows is which items tend to be grouped together by these subjects. The most important groupings come first.

The first-order factor for the males and the first-order factor for the females made an interesting contrast (both loading the major part of the overall variance: 35 percent). The male factor was a combination of high-power tactics: physical power (attack and restraint), strategy (flattery, making the other person feel guilty), and arbitrary power (command, reprimand), which is the sort that goes with ascribed rather than achieved status (and that French and Raven [1959] call legitimate power). This combination fitted stereotypes about executive heads of government and military generals who support their power by charisma, strategy, and a big stick. In the sibling data, firstborn of both sexes gave themselves and were given by their siblings higher scores on all of the items of this generalized power factor. In an earlier study with children, it was found that the most successful children were also generally seen by their peers as good at both the physical and intellectual manifestations of power (Sutton-Smith & Roberts, 1964). It will be noted from the table that the female second-order factor is of the same general character as this first-order male factor, though in the case of the females the strategic artifices are more numerous (as in giving rewards, making

---

* Varimax rotation of the first six factors was used, as there was little subsequent difference in the latent root.

TABLE 4.8 A Factor Analysis of Power Tactics Used by Siblings·

Self to Sibling—College Males (N = 120)

 I. Command (.78); Make feel guilty (.67); Physical restraint (.54); Flattery (.50); Reprimand–scold (.45); Physical attack (.42); Plead and whine (−.57).
 II. Sulk (.69); Appeal for sympathy (.67); Ask because a girl (.65); Stubborn (.60); Not be friendly (.55); Plead and whine (.44); Angry–shout (.42); Wishful thinking (.31).
 III. Ask other kids for help (.73); Ask because a boy (.72); Reprimand–scold (.47); Request (−.60); Reason–explain (−.35).
 IV. Persuasion (.70); Give rewards (.70); Blackmail (.69); Bargain (.68); Bribery (.57); Appeal to reason (.42); Request (.35); Reason–explain (.34).
 V. Ask Dad for help (.66); Ask Mom for help (.66); Appeal to reason (.60); Ask because younger (.59); Glorify thing to be done (.58); Make feel obligated (.52); Plead and whine (.42); Prayer (.38); Reason–explain (.35); Stubborn (−.33).
 VI. Threaten (.70); Harass (.63), Ask because older (.61); Angry–shout (.61); Physical attack (.55); Physical restraint (.54); Teasing (.51); Trickery (.47−); Blackmail (.39); Bargain (.35); Flattery (−.37); Reason–explain (−.32).

Self to Sibling—College Females (N = 120)

 I. Threaten (.72); Sulk (.72); Appeal for sympathy (.71); Angry–shout (.68); Stubborn (.68); Teasing (.61); Trickery (.52); Not be friendly (.50); Harass (.50); Plead and whine (.49); Make feel guilty (.37); Bribery (.34); Physical attack (.34); Ask other kids for help (−.43); Appeal to reason (−.35).
 II. Command (.68); Physical restraint (.69); Reprimand–scold (.67); Flattery (.60); Give rewards (.47); Ask because younger (.45); Make feel obligated (.43); Ask because a girl (.38); Appeal to reason (.37); Glorify (.32).
 III. Reason–explain (.66); Request (.60); Persuasion (·.58); Bargain (.56); Glorify thing to be done (.51); Appeal to reason (.47); Ask because older (.47); Wishful thinking (.27).
 IV. Appeal to reason (.43); Ask Mom for help (−.77); Ask Dad for help (−.77); Harass (−.46); Ask because a girl (−.40); Plead and whine (−.36).
 V. Bribery (−.77); Blackmail (−.62); Give rewards (−.50); Trickery (−.46); Bargain (−.45); Make feel obligated (−.39); Flattery (−.33); Plead and whine (.25).
 VI. Physical attack (.58); Wishful thinking (−.59); Prayer (−.58); Appeal for sympathy (−.34); Flattery (−.32); Physical restraint (.38).

SOURCE: B. Sutton-Smith and B. G. Rosenberg, "A Factor Analysis of Power Styles in the Family." Paper presented at the American Psychological Association, New York, September 1966.

the sibling feel obligated, appealing to reason, and glorifying the job to be done). In earlier research, we had found also that females play more strategy games than males do (see Roberts & Sutton-Smith, 1962).

The first-order factor for females seemed to be the most exciting discovery in the present study, but it is difficult to characterize it without succumbing to masculine prejudice. In previous analyses of this scale, these were all items which had low social desirability, and which were more likely to be used by sisters against brothers (MF2, F1M) than by sisters against sisters (FF2 and F1F). The negative loading on rationality (appeal to reason; reason–explain) and the apparent irrationality of the other variegated devices are noticeable. Speculatively, the items might represent moments in a series which begins perhaps with appeals for sympathy; moves to pleading, whining, and sulking; shifts to threatening, not being friendly, and being stubborn; switches again to teasing, harassing, and trickery, making others feel guilty, and bribery; and ends by physical attack and shouting. Perhaps we ought to think of this series of response factors as having a temporal dimension—that is, made of consistent changes over time under stress. It is a difficult power factor to name, but some female literary figures (such as Scarlet O'Hara and Jezebel) seem to fulfill it. That other source of cluster analysis, the Thesaurus, puts together a group of feminine traits which have a somewhat similar ring; namely feline, tigress, tart, bitch, harridan, strumpet, minx, and slut. We have preferred to call this the *trollop factor*. Whatever is the appropriate name for this clustering of supplication—negativism, trickery, and anger—it is a discovery worthy of further systematic work. In previous work with females, attempts to apply the notions of power derived from games were certainly not particularly successful, perhaps because games do not model any such trollop factor (Sutton-Smith & Roberts, 1964).

The second factor for males was a general low-power factor, not unlike the girls' first-order factor but having less strength. The female strategic elements of threatening, teasing, trickery, harassing, and bribery were missing from the boys' factor as was physical attack. In earlier work these low-power items of supplication (sulking, appeal for sympathy) and negativism (not be friendly, stubborn), and anger were more often associated with second-born boys.

Factors V for males and III for females were clearly strategy factors —though there were differences between the sexes (by "strategy" is meant here the use of intellectual means of control). The girls' strategy factor contains items of predominantly high social desirability (reason–explain, persuasion, request, and appeal to reason), whereas the boys' factor has these and adds the strategic items of low social desirability (blackmail, bribery, trickery). If these two are thought of as positive and negative components of strategy, it is noticeable than that the boys have combined both components in the one factor (IV), but the girls have the negative components only in their first-order trollop factor. A further finding is that firstborn boys have higher scores than non-firstborn boys on all these items. The rather feeble character of purely positive strategy for the girls is suggested by the fact that wishful thinking gets its highest positive loadings (.27) on this female strategy factory (III) and on the boys' low-power factor (II) (.31).

The nearest things to physical power factors are numbers VI for both males and females. Again, the female physical factor is comparatively simple. The only positive items are physical attack and restraint, and the negative items are a collection of very low-power items of a magical sort such as prayer and wishful thinking. The male physical factor, however, is more like the female first-order factor, though without its weak supplicatory and negativistic elements. For males these are all low social desirability items emphasizing physical modes and negative strategy—a sort of gangster factor it might be called.

Factor III for boys involved dependency on others and negative loadings on positive strategy; a peer factor, perhaps.

Factor V for males involved dependency on parents by supplication, together with positive strategy and magic. Again, there is the association between positive strategy and magic (as in Factor III for girls), though with the addition of dependency in the case of the boys. This is an interesting finding because in earlier work, the influence of chance or fortune or magic has been hard to track down in middle-class subjects, who ordinarily deny such influences. These associations might imply that among these middle-class children the use of sweet reason (reason, explain, persuade, and appeal to reason) is tantamount to wishful thinking and prayer. Simple reasonableness is equated with a dependence upon fortune!

The other female factors are characterized more by their negative than by their positive loadings, as if agreement were more about what not to do than what to do—a finding which replicates some earlier work on recreation differences between the sexes, in which the females were more often easier to characterize by avoidances than by interests (Phelps & Horrocks, 1958). Thus, Factor IV for girls has loadings on positive strategy but more negative loadings on supplication and negative strategy. And Factor V for girls is negative on all types of strategy, its only near positive factor being supplication, i.e., plead and whine (.25), and Factor VI is negative on wishful thinking and prayer while emphasizing physical tactics. It is difficult to be sure of what all this consensus of a negative sort means, except that in the earlier study we also showed that girls make more meaningful discriminations of types of failure than boys (Sutton-Smith & Roberts, 1964). Perhaps negativism and avoidance can be regarded as a type of power.

Unfortunately the number of subjects available in each of the ordinal categories was not sufficient to permit a factor analysis by ordinal category, though some of the probable associations have been inferred throughout the foregoing discussion. The factor analysis does make clear, however, that it would be worthwhile to undertake an observational study of power tactics based on the categories given above. There is apparently a wealth to be learned about the ways in which children grow wise in the use of "power." However, we have generally avoided such a study because of our inhibitions with respect to the direct study of such matters either in adults or children (French & Raven, 1959). Clearly, children are learning somewhere how to lie, deceive, blackmail, harass, and bribe, and the most likely situations are their games (Sutton-Smith & Roberts, 1964), their peer activities, and their "naked" lives with their siblings. No one can be wise in war, politics, or marriage without knowing how to tell, at the very least, "white lies." The game begins presumably with brothers and sisters.

AGGRESSION AND ANGER

We move now to review the literature on sibling differences in aggression and anger. Most of these materials take on a clearer light in terms of the analysis of power differences given above. All the evidence suggests that the later born are more affected by their older siblings than vice versa. But the question that is difficult to answer, with respect to power relationships, is whether the younger sibling's behavior can be classed as a reaction to the pressure of the older siblings (anger and sulking, for instance, certainly fit this category), or whether it is largely modeled after the rather crude dominance of the older siblings. The younger siblings' power assertions might be expected to be less controlled than those exerted by the firstborn, who might be modeling after parents rather than siblings. There is probably not a general answer to these questions. Adler (1959) suggested that later born become power seekers as a result of their long years of subjection. This would be, in our terms, a predominantly modeling thesis. Irving Harris (1964), pointing to Hobbes and Machiavelli as second born, also seems to suggest that the second born have a greater interest in revolutionary power than the firstborn (who take their own messianic powers for granted). However, empirical work is scant. There are some indications that second born are more inclined than firstborn to the exercise of power over others (Veroff, 1957; Krout, 1939; Singer, 1964). Second born claim to be more often the boss in their play with friends. Yet in a sociometric study conducted by the present investigators (Sutton-Smith & Rosenberg, 1968), they were *not* more often seen as bossy by their class peers. If their bossing attempts were of a crude character, one would expect them to be less popular, but the evidence at the elementary school level has proved otherwise. In fact, evidence for later borns' popularity leads to the conclusion that they do not model simplemindedly after their overbearing firstborn elders, but that they must have learned a directness and sociability, and perhaps "Machiavellianism," in dealing with those elders, which transfers elsewhere. The evidence that "Machiavellianism" operates more often in face-to-face peer contacts, where there is latitude for improvisation, and where irrelevant affect must be controlled, certainly appears to favor the later born (Christie, 1967). Still, this must be a hypothetical position until there is more evidence about ordinal differences in the application and development of power tactics outside the family setting.

Whatever may be the conclusion with respect to the way in which the second born model after the power of older siblings (or develop tactics in relationship to them), there is some information about the way in which they react to that power, and attempt to counteract it. In Chapter 3 it was noted that under strong opposite-sex role influence, males, in some conditions, have counteracted it by an equally extreme adherence to masculine preferences or activities. Do the second born under the stress of firstborn power show some analogous form of counteraction? Their direct response as reported above was to get angry, sulk, appeal for help, or isolate themselves. Are they in general made angrier and more aggressive as a result of this continuing power subordination?

Several studies at the preschool level have established that the later born are indeed more overtly aggressive. Gerwitz's work (1948) will be mentioned in the following chapter on affiliation. In his study, negative attention-getting became more prominent in the later-born positions. Pauline Sears (1951), working with preschool children, similarly found later borns to be more aggressive. She observed 136 three- to five-year-olds enrolled at day care centers for children of working mothers in three cities in Iowa and Illinois. The subjects were of lower-middle and upper-lower social status. Each child was given two 20-minute periods of play on different days with the doll play equipment—six rooms with furniture, open at top, and with toys in it (mother, father, brother, sister, and baby). The child was told: "You can make them do anything you want. You go ahead and play with them anyway you like," While the child played, he was encouraged by the examiner in whatever he did. The examiner scored the play for the frequency of acts of aggression and the nature of these acts (scolding, fighting, and so forth). The salient finding was that younger children of both sexes showed more aggression than older children, and that only children were like younger children in being more aggressive. Not only is this finding similar to that of Gewirtz, it is also similar to that of other studies. Goodenough and Leahy observed and rated older siblings on an aggression scale as less aggressive than only children and younger siblings. Of similar import is Nisbett's finding (1968) that later borns participate more often in dangerous sports. According to Nisbett they are less frightened by the prospect of physical injury. Their superiority as fighter pilots and aquanauts (Chapter 5) is of similar significance. These findings are an important complement to those of Sears because they indicate that her later borns' aggressive doll play fantasies are an expression of general behaviors, not solely a compensation for their inability to express aggression elsewhere, though that might also be true. It could be interpreted, for example, that only children and younger children get less scope for retaliation in the home.

Some of Pauline Sears' more detailed findings (1951) are worth mentioning, given the consistencies across the studies mentioned above. Older sisters showed practically no aggression at all. Older brothers' scores were like the girls' mean scores. Both older brothers and sisters were alike in avoiding the rough, injurious, hurting types of aggression which were typical of younger siblings. This difference importantly parallels the types of problem behavior typical of these children to be cited in MacFarlane's study shortly—with the older child internalizing more and the younger child externalizing more. In the doll play stories, older boys had the father exhibiting and receiving most of the aggression, whereas older girls had the girl and the baby giving aggression, and the mother, the object of it. Each apparently focuses more hostility on the same-sex parent. Younger brothers had the mother giving most of the aggression and the mother getting most of it, an unusual finding but perhaps consistent with the statements of the mothers in Sears, Maccoby, and Levin (1957), who appeared to feel coolest of all about second-born boys, particularly if they had already had a boy. Only boys saw the mother as both giving and getting most of the hostility. This difference is somewhat similar to that shown by firstborn boys and girls, except that levels of aggression are much more pronounced in the case of the only children. Again, this is interesting because in other studies the cross-sex identification of the only children is much more pronounced than that of the firstborn. That is, female only children appeared to identify with fathers, and male only children with mothers.

MacFarlane, Allen, and Honzik's data on children's problems also seem to imply that the later born strike out more directly, are less inhibited, and externalize more. In "A Developmental Study of the Behavior Problems of Normal Children between Twenty-one Months and Fourteen Years," MacFarlane, Allen, and Honzik (1954) presented the mothers' reports on the problem behavior of their 104 children, year by year, during the children's growth over those 14 years. Table 4.9 shows the

TABLE 4.9 Problem Behaviors and Ordinal Position

Boys

Firstborn		Non-firstborn	
Oversensitive	7 years	Irritability	21 months
Restlessness in sleep	7 and 12 years	Jealousy	21 months
Disturbing dreams	9 and 12 years	Destructiveness	4 years
Physical timidity	9 years	Food finickiness	7 years
Excessive demand- ing attention	9 years	Excessive competitiveness	8 years
		Speech problems	10 years
		Overactivity	11 years
		Temper tantrums	13 years

Girls

Excessive demand- ing attention	3 and 8 years	Specific fears	21 months
Overactivity	5 and 6 years	Oversensitive	4 years
Tics and mannerisms	6 years		
Irritability	6 and 11 years		
Shyness	7 and 11 years		
Restlessness in sleep	8, 10 and12 years		
Somberness	8, 9 and 11 years		
Temper tantrums	8 and 11 years		
Nail biting	9 years		
Lieing	9 years		
Excessive reserve	9 years		
Physical timidity	11 years		
Mood swings	11 and 12 years		
Negativism	11 and 12 years		

SOURCE: Adapted from J. W. MacFarlane, L. Allen, and M. P. Honzik, A Developmental Study of the Behavior Problems of Normal Children between Twenty-one Months and Fourteen Years, *University of California Publications in Child Development*, 1954, **2**.

results of comparing firstborn with non-firstborn for significant differences in behavior at different age levels. MacFarlane et al. say:

> The data herein are from mothers. It is conceivable and even probable that somewhat different frequencies would have been obtained had we secured data from the fathers. Further, the data may have a systematic error in reported frequencies for girls and for boys. It is possible that a mother may have a higher expectancy of "good" behavior from her daughter and thereby be more critical of certain of her lapses from expectancy. It is possible, too, that energetic aggressive male behavior may be irritating and freely reported. Similarly, fathers might be more critical of certain of their sons' behaviors and more accepting of aggressive be-

havior. All we can say for certain is that the frequencies presented in this paper are colored by the mother's perceptions, and reports by her probably represent a fairly true picture of the mother-child relationship, an important part of the atmosphere in which the child develops his patterns" [1954, p. 4].

The most obvious difference according to MacFarlane et al. is that the firstborn characteristic problem behaviors are more often of an "internalizing" character, while those of the second born are more "externalizing." These reports are consistent with the studies of P. Sears (1951) and Gewirtz (1948), mentioned above, showing the non-firstborn to be more aggressive. A second point of some interest, however, is that the problems of the firstborn are, with one exception, reported only after the preschool period (1/32), while half the non-firstborns' problems are reported for the preschool period (5/10). We might perhaps conclude that while the probability of problem behavior remains fairly constant over this period for the non-firstborn, it increases markedly with the commencement of schooling for the firstborn. This is a finding which prompts the digression that the status of the firstborn during the first five years is a less adequate preparation for the transfer to secondary ties than the less-preferred status of the later born. The important point at present is that in all these studies in the early school years, we have indications of greater aggressiveness on the part of the second-born males.

Are these to be accounted for in the patregenic terms (that is, accounted for by parental influence) that Gewirtz adopts, or are they an outcome of being put down by the older, more powerful siblings? MacFarlane et al. (1954) suggest:

> In a competitive culture, especially for the males, the second born male is in an inferior and deflating position . . . in size and strength . . . and being a male, strikes back overtly and competitively. . . . [he] feels inferior about ability and constantly, in counteractive fashion, strives to equalize matters. . . . [The firstborn] was not allowed physical retaliation against a less strong, younger sib, and was "hurt," that is, managed his strains by internalization (in dreams, over-sensitiveness, mood swing, fears) and by indirect deflation of a younger sib in the sib's area of vulnerability, namely, with such verbalizations as "you're dumb" or, in a long suffering voice, "here stupid, let me do it!" [p. 181].

Sears, Maccoby, and Levin (1957), who began their own work also with patregenic assumptions, were forced to the view that aggression among children was better explained in "social structural" terms. That is, they had supposed that adult permissiveness for aggression would be systematically related to the amount of aggression shown in the different ordinal positions. They found, however, that these two variables were not related. They therefore suggested instead a social-structural view as follows: " . . . relatively greater amounts of frustration and discomforting control in a family comes from the persons who are immediately above the child in the power hierarchy than from other family members, and regardless of the parents' permissiveness and punitiveness, the younger child tends to be more aggressive towards those persons" (1957, p. 418). We might argue that the later born may simply be more aggressive in their nursery play because that is the way they have learned to relate to others, from their experiences with their older siblings.

In sum, these data seem to show that later-born children counteract the power of the older children by being fairly directly aggressive and externalizing in their responses. They may not actually be, in general, any more hostile than the firstborn, but they do appear to express themselves in a more spontaneous and primitive fashion. Firstborn tend to be more aggressive in an adult way, deflating the younger born with verbal criticism and creating alibis for their own inadequacies (Koch, 1955b).

CONCLUSION

In this and the previous chapter, dealing, in particular, with child-child interactions, the data appear to show the very clear effects of sibling sex status and sibling power. Females are typically more affected by males than are males by females. Second born are influenced by the characteristics of their older siblings. The data seem uniform in the case of females. But males in some circumstances seem to counteract the influence of the females by asserting their own sex-role characteristics. Firstborn and second-born siblings act like members of a dominance hierarchy, with the firstborn showing the higher power tactics, and the second born showing many forms of counterreaction, including aggression, against these power tactics. The way in which the data appear to be congruent with this social-structural framework (an explanation in terms of a dominance hierarchy) would appear to be the best available evidence for the view that regardless of parents, the sibling-sibling interactions are intrinsically responsible for many of the established sibling behavior and personality differences.

academic primogeniture

The previous chapters have emphasized the influence of siblings upon each other. This emphasis has been dealt with first because it is a particular burden of this book to counteract too many years of overemphasis on the importance of parent-child influences alone. This is not meant, however, to deny the importance of parent-child influences, nor to suggest that child-child influences are as powerful. It is conceivable that child-child influences are as important as parent-child influences are with later-born children, but it is not conceivable that they are as important with firstborn and only children. This chapter and the next two chapters, therefore, focus on the special accomplishments and character of firstborn and only children. After their special relationships with parents have been investigated, we shall, in the final chapters, turn to the questions of the way in which those relationships and the relationships discussed in Chapters 3 and 4 may be fitted together into the one pattern of family interaction.

It has long been asserted that the firstborn are more intelligent, eminent, scholarly, and achieving. The soundness of these assertions will be reviewed before proceeding further.

INTELLIGENCE

There is no genetic reason for expecting intelligence tests to reflect differences in birth order. After his most careful survey of studies purporting to demonstrate such differences, Jones (1931) says: "We prefer to conclude that intelligence is not yet proved to be a variant with order of birth. Such a negative finding is in accord with expectation if individual differences in intelligence are determined chiefly by genetic rather than epigenetic factors" (pp. 226–227). Several of the most superior studies involved very large numbers and did appear to show results giving a slight advantage for the later born. Jones (1931) was able to demonstrate, however, that these birth-order differences were artifacts of scale construction. Through defects in the standardization and scale construction of the tests used at that time, all subjects tended to get lower scores with increasing age. Therefore, since the firstborn subjects when compared with their own siblings were inevitably older, they appeared to have smaller scores as a group than their younger siblings. There was a negative correlation between chronological age and IQ of about −.3.

Although various writers in the recent literature have gone on to discover differences between birth order and IQ, there does not appear to be any reason for modifying Jones's verdict. It is quite possible to get differences if any or some of the methodological controls mentioned in Chapter 2 are not carefully observed. For example, when firstborn in general are compared with later born, members of large families are overrepresented in the later-born group. This follows because each large family contributes many subjects to the later-born group but only one to the firstborn group. Contrarily, children from small families contribute more equally to both firstborn and later born, though only perfectly so in the case of two-child families. As different size families themselves vary in many characteristics—sometimes in socioeconomic status, in intelligence test scores, in parental enjoyment in having children, in the value similarity of parents and children, and even in physical abnormalities—it is quite possible for these differences to be mistaken for differences resulting from ordinal position. (Strodtbeck & Creelan, 1968) For example, Jones (1931) suggests that the occurrence of abnormality in early-born children tends to lead to a restriction on having more children, with the result that small families are pathologically weighted, and there appears as a result greater relative physical abnormality among the firstborn (p. 208). By and large, more of the studies since Jones have controlled for family size than had previously been the case. The most adequate method of controlling for family size is by contrasting different ordinal positions *within* families of different sizes: first versus second in two-child families; first versus second versus third in three-child families.

Still, although there is no good reason for assuming genetic differences between ordinal positions, the experience of birth itself may differentiate between them. It may be more anxiety-arousing for firstborn (Weller & Bell, 1965; Weiss, 1967a). There is some tentative evidence for the view that the early born are more variable in intelligence. It has been suggested that the youth of the mother may have, alternatively, a beneficial physical effect, or, alternatively, a negative effect owing to difficulties in delivery (Bayer, 1967). But findings in this area are too uncertain as yet for any positive verdict.

EMINENCE

A long line of investigators including Galton, Ellis, Cattell, Gini, Clark, Huntington, Apperly, Jones, and Roe have reported that there are more firstborn amongst great scientists, eminent people, professors, physicians, men of letters, Rhodes Scholars, and those listed in *Who's Who* (Schachter, 1963). Unlike the studies of intelligence, few counterclaims have been made for the later born. Altus (1966a) quotes a communication by Nichol on National Merit Scholars (the winners in a national exam for exceptionally able students) showing that in all-size families firstborns are overrepresented. But this overrepresentation occurs, he says, only among those with extremely high scores (good enough to be National Merit Scholars). In the large body of students who took the tests in the first place, there was no general relationship between ordinal position and scores. Similarly, Terman (1925), in his studies of children of highly superior intelligence, found that the firstborn were overrepresented. All his children had an IQ over 140 and were, in effect, the brightest 1 percent of children in and about the San Francisco Bay Area. Apart from the fact that there was no overrepresentation of the youngest in the National Merit Scholars mentioned above, these two studies seem to concur. That is, while there is no general relationship between IQ and birth order, and between National Merit performance and birth order, if the distribution is cut off at the top, the very superior persons do evidence such birth-order differences. There is other support for this relationship (Sheldon, 1954), and even evidence that our astronauts are either only or firstborn children (*New York Times*, December 24, 1968). Altus (1966a) concludes: "In England and the United States there appears to be an indubitable relationship of birth order to the achievement of eminence, however it has been defined. The dice are loaded in favor of the firstborn. . . . Here is intellectual primogeniture with a vengeance" (pp. 2, 5).

Despite the tenacity of this tradition that there is some special relationship between being firstborn and being eminent, a number of sceptical voices have been raised both against it, and against the simplicity of Altus' conclusion (Datta, 1967; Lunneborg, 1968). The most substantial critic has been Bayer, although in a survey of some 8,000 persons receiving the doctorate in 1962, his conclusions only modify those of Altus without actually contradicting them. Bayer (1967) says: "In each sibship size from two to five, a greater number of doctorates are first born than is expected. Those in an intermediate sibship position are consistently under-represented. Youngest, however are under-represented only in two and three child families; in four and five child families a slightly greater than expected proportion are last born" (p. 546).

Various explanations are offered to support these differences. The simplest is that of Schachter, who suggests that more firstborn are eminent because more of them get to college in the first place.

COLLEGE ENTRANCE

The data, here, like the data on eminence, appear to show that first-born are overrepresented in college, including graduate programs, and that the degree of overrepresentation increases with the level of education involved.The more highly selective the college, the greater the degree of over representation, varying from over 60 percent in the famous private colleges (Yale, Reed) to around 50 percent in large state colleges compared to some 30 percent to 40 percent of firstborn in the general population. Schachter suggests, therefore, that there is a relationship between the eminence mentioned above and going to college.

> The domains in which fame has been assayed almost all require scholar-ship and education. It is conceivable that first borns are simply more bookish or that their educational opportunities are greater. If true, the consistent over-representation of first borns among eminent scholars may reflect nothing more than an over-representation of firstborn among all scholars—eminent or not [1963, p. 759].

Before we accept Schachter's interesting suggestion, however, some demographic precautions are in order. For even when family size or socioeconomic status is held constant, there is still the possibility that an overrepresentation in the groups we are studying is due to some special characteristic of the larger population of which it is a part, such as an accelerating birth rate or a trend toward younger marriages. In some new suburban areas, for example, with younger parents, for some time the schools of the neighborhood will have an overrepresentation of early-born children. The later born have not yet arrived in the school sample. This was the case in an earlier study of these investigators (Rosenberg & Sutton-Smith, 1964b) among 900 preadolescent children in northwestern Ohio, grades 4 through 6. This study, which included all children present at these schools, yielded the following numbers of children in three-child families:

Firstborn boys: 39   Firstborn girls: 48
Second-born boys: 33   Second-born girls: 43
Third-born boys: 19   Third-born girls: 15

If firstborn are indeed different from later born, then this school should be a very different place several years hence when firstborn cease to exercise such a disproportionate influence on the upper reaches of the school. A more important source of variation (than geographic difference in the youth of the population at large) is historical variation in birth rates. For example, the analysis of nationwide nativity statistics, by Bayer (1966) shows that the proportions of birth-order positions born in any given year have changed considerably over the past 40 years. During the 1920's the percentage of firstborn appears to have been in the lower 30's. This percentage increased gradually toward a peak of approximately 43 percent in 1942, stayed high until about 1947, and then began a gradual decrease until it was only 27 percent in the 1960 figures. A sample percentage distribution from Bayer's compilations is shown in Table 5.1.

TABLE 5.1 A Sample Percentage Distribution

Year	1st	2nd	3rd	4th	5th	6th and Later
1960	27	26	20	13	7	8
1947	43	28	14	7	4	6

Source: Adapted from A. E. Bayer, "Birth Order and College Attendance," *Journal of Marriage and Family Living*, 1966, 28, 480–484.

These are rounded percentages and, at best, approximations to the real phenomena because of differences in registration, sampling, and calculating procedures. But the effects of such variations on any attempts to judge whether firstborn or others are overrepresented among the eminent or in college populations are clear. Judgments of overrepresentation are usually made on the basis of a current sampling from a larger population, say, a high school population (Schachter, 1963), whereas the more correct sampling is probably the total population at the time of the subjects' birth.

These strictures of Bayer cast some doubt on the apparently overwhelming evidence of the relationship between first birth order and eminence in general (1966, 1967). Making the same sorts of observations as those made by Schachter, but also taking into account the vitality statistics, Bayer shows that in a large national sample of 18-year-olds who were born in the year 1942, when 43 percent of all births were firstborn, 48 percent at the senior high school level were firstborn and a year later among those attending college, 55 percent were firstborn. When controls were introduced for family size, Bayer again found the same disproportion of firstborn, though it became more marked as the family size increased. Still Bayer continued sceptically:

> Changes and fluctuations in the marriage rate, age at marriage, completed family size, age of mother at first and last births, spacing of children, age structure of the population may all affect the proportion in a given ordinal position in a sample at any point in time. . . . [Only a] comparison of proportions of college and non-college goers *within* family sizes *and* birth positions provides a control for this complex of demographic changes which may otherwise yield spurious data [1968, pp. 483–484].

Table 5.2 from Bayer presents such a comparison: percent of students in senior college one year after high school, by ordinal position, sex, and socioeconomic status. It demonstrates that last born are about as likely to go to college as firstborn from the same family size and socioeconomic

TABLE 5.2 Percent of Students in Senior College One Year After High School, by Ordinal Position, Sex, and Socioeconomic Status

			Ordinal Position					
	Only Child (1 of 1)		Firstborn (1 of 2, 3, 4, 5)		Intermediate Born (2 of 3; 2, 3 of 4; 2, 3, 4 of 5)		Last Born (2 of 2, 3 of 3, 4 of 4, 5 of 5)	
Sex and Socioeconomic Status	Total Weighted N (in 100s)	Percent in Senior College	Total Weighted N (in 100s)	Percent in Senior College	Total Weighted N (in 100s)	Percent in Senior College	Total Weighted N (in 100s)	Percent in Senior College
Male:								
Low SES	105	21.7	460	18.9	323	18.5	305	18.9
Low-middle SES	162	46.5	588	35.5	356	24.2	290	32.2
High-middle SES	134	55.3	587	48.7	283	46.1	253	52.8
High SES	120	74.8	596	70.5	229	63.2	270	71.5
Total	521	50.3	2231	44.9	1191	35.4	1118	42.7
Female:								
Low SES	139	16.3	519	11.9	431	10.7	304	13.1
Low-middle SES	183	25.9	719	23.1	495	15.8	366	21.8
High-middle SES	175	44.9	686	40.2	342	29.5	321	43.0
High SES	129	64.9	557	62.8	232	58.6	242	59.1
Total	626	37.0	2481	34.3	1500	24.1	1233	32.5

SOURCE: A. E. Bayer, "Birth Orders and College Attendance," *Journal of Marriage and Family Living*, 1966, **40**, 480–484.

position. Further, within each socioeconomic level only children are more likely to attend college than are those of any other ordinal position. Least likely to get to college are those of an intermediate family position. It is clear that any gross comparison of firstborn, particularly if it includes only children, but even without including them, will appear to favor firstborn over all later born. Furthermore, the high percentages of college attenders among the last born will be reduced by averaging them in with the intermediately born. This is an interesting result because it does not negate earlier firstborn versus later-born contrasts, but it does indicate that the more basic problem concerns the special position of the underrepresented middle-born children. In most studies firstborn and only children are not separated, so that we cannot know to what extent each contributes to the outcome. For most of the data in the literature firstborn must be considered as a composite group of first-born and only children. What must be explained, it seems, is the particular position of the only children and the last born, or, alternatively, the unfortunate position of the middle born.

ACHIEVEMENT

Most students of birth order and eminence have not assumed that eminence is merely an artifact of college attendance. Some have suggested that eminence is due to differences in intelligence, perhaps underlain by the differential physical (anxiety-inducing) effects of birth. The above-mentioned difference between only children and firstborn would not support such an explanation. As far as birth is concerned, they have the same firstborn experience, whatever that may be. Others have suggested that the firstborn is given an economic advantage. He has prior access to the family resources. This explanation, if tenable, would have to require that the family's resources recover again by the advent of the last born. Most recent thought on this matter, however, has favored explanation not in terms of physical or economic variables but in terms of socialization differences between the various ordinal positions. The simplest socialization explanation would probably be that relative to all others the middle born are neglected. They do not get the exclusive attention of being firstborn, or the doting attention of being youngest (at least not for long). Practically every evidence that will be cited concerning middle borns throughout this book has a negative implication. They show more negative attention-getting (Gewirtz, 1948), are most changeable (Brock & Becker 1965), are less often given affectionate nicknames by parents (Clausen, 1966), and are least popular (Sells & Roff, 1963). In years of research on parent-child relationships these are the types of responses that have most often been associated with neglect or harsh treatment. Conversely, it would not be difficult to argue that those who get relatively greater care and attention from parents (the only children most of all) exhibit the types of responses that parents usually encourage. So it would follow that the middle born would do less well at those things most conventionally encouraged by adults such as school achievement, and that the others, in particular the only children, would do the best of all.

Most of the research on birth order and achievement, however, has not taken this direction, but has assumed that it was the special case of the firstborn that required explanation. In the rest of this chapter, therefore, we pursue this more general trend in the research, keeping in mind that it is, in fact, the contrary poorer standing of the middle born which is the empirical state most in need of explanation. The demonstration of firstborn superiority has taken two forms: (1) by showing that the school grades of the early born are superior, and (2) by demonstrating that they have higher need for achievement as this is assessed by various psychological tests of achievement motivation. When one considers the pitfalls that lie in the path of attempts to demonstrate birth-order–intelligence relationships, one is reluctant, understandably, to rely on the handful of studies that currently indicate either that firstborn do better at school or that they have higher need achievement. Yet such studies as we have of school grades do seem to indicate a superiority for the firstborn (Pierce, 1959; Elder, 1962; Konig, 1963; Schachter, 1963; Oberlander & Jenkins, 1967; Chittenden, Foan, Zweil, & Smith, 1968). The number of studies is small, but there are as yet no strong counterindications.

If students who are firstborn get better grades at school, which is the reason for their going to college and becoming disproportionately eminent, then why do they get better grades? From the fact that good grades have been found to be a good predictor of motivation for academic success at college, we might assume that firstborn get good grades because they are motivated to do so; that the grades not only facilitate college entry but are an index of scholastic motivation. For example, Elder (1962) found relationships between high school grades and a student's self-report. The student with good grades was interested in his schoolwork, he tried hard to get good grades, he studied at home, he finished his homework, and he tried hard to improve when not doing well (p. 43). In other words, the grades and motivation went together.

If the desire for achievement is measured by some of the better-known projective tests in the psychological literature, however, the results are equivocal. Some results using projective tests have been positive (Sampson, 1962; Bartlett & Smith, 1968; Sinha, 1967). Sampson and Hancock (1967) had partially positive results using the Edwards Personal Preference Schedule (EPPS), which is a verbal inventory. They distinguished the eight ordinal positions as well as the only children, and the rank order of achievement responses across these positions was

M1F–M–MM2–M1M–F–F1F–FM2–MF2–F1M–FF2. This is almost the reverse order of that presented for affiliation in the following chapter. The same EPPS scale apparently produces a strong negative relationship between affiliation and achievement across these ordinal positions. Here we find that the two-child males (MM2) who scored least affiliative, scored high on achievement and vice versa; the females (FF2) who scored most affiliative, scored low on achievement. One partially corroborative study is that carried out by Teepen (1963), in which she contrasted boys with older brothers against boys with older sisters, and found that on an achievement measure derived from the California Psychological Inventory the boys with older brothers made more achievement responses. In Sampson's rank order (1962) mentioned above, the MM2 are also clearly ahead of the FM2. Projective studies by Moore (1964) and Wolkon and Levinger (1965), however, have had negative results. Rosenfeld, in a comprehensive study of five different samples in which he varied the sex of the stimulus figures (for a McClelland type of test of T.A.T. n Achievement), the group or individual character of the testing, the degree of arousal, and the sex of the experimenter, found that these conditions interacted in various ways with the subjects' responses, so that it became most difficult to draw a conclusion with respect to motivation simply by administering such a projective instrument. We may perhaps conclude not that higher achievement motivation among the firstborn has not been supported, but only that projective and self-report measures are so susceptible to a host of varying influences that they cannot be soundly used to give a verdict on the issue. The clearest evidence that during the course of development the firstborn are more achievement oriented comes from the interview studies of 122 mothers and their 122 eleven- to fourteen-year-old sons by Rosen (1964) to determine to what extent they agreed with each other in their values about achievement. Did both mothers and sons feel that the child must strive to achieve, or did they both feel that personal achievement is impossible and pretty much a matter of luck? Results were that only children and oldest children had achievement values most similar to those of the mother and that the middle born and the youngest had values less like those of the mother, but this was found only in the higher socioeconomic levels. At these higher socioeconomic levels the mothers said they trained the children earlier and used more "conditional love" disciplinary techniques. At all class levels, small and medium-size families had higher levels of value similarity than large families (Rosen, 1964). In an earlier report of the same research by Rosen (1961) in which the boys were tested with an n Achievement projective device, it was also found that boys from smaller and medium-size families tended to have higher need achievement than boys from larger-size families. But birth order interacted in a complex way with family size and socioeconomic level. There was no difference between ordinal positions in the two-child higher status families. They both had high need achievement. But in the medium-size and large-size higher status families, the firstborn did show higher need achievement than the youngest. In the lower classes, in all-size families, the youngest was more likely to show higher need achievement. Table 5.3 indicates some of the complexities. The results are not unsimilar to those of Bayer presented in Table 5.2 above.

TABLE 5.3 N Achievement

	Social Class I–II–III			Social Class IV–V		
	Family Size			Family Size		
Birth Order	Small	Medium	Large	Small	Medium	Large
Oldest	5.82	7.52	5.75	4.31	2.86	1.00
Intermediate	*	5.44	10.00	*	3.43	1.96
Youngest	5.94	5.21	2.00	5.93	3.90	2.84

Source: B. C. Rosen, "Family structure and achievement motivation," *American Sociological Review*, 1961, **28**, 274–585.
 * There are, of course, no intermediate children in a two-child family.

In these lower-class families, the older boys by and large were reared by young mothers with whom they had low similarity in values. Rosen suggests that these young mothers were impatient and perhaps resentful of the pressure put on slender family resources. We might suppose also that these elder boys were forced to play surrogate roles, and wherever possible to contribute toward the economic resources—a situation that Elder (1962) indicates is deleterious to their school performance. The remarkably high scores of the middle children in the upper-status large families is worthy of note because it contradicts all other information about the middle borns. It defies easy explanation.

Recollecting Bayer's finding concerning the disproportionately large numbers of only children at college, it is interesting to note that in Morris Rosenberg's large-scale study of self-esteem (1965) the only children were also distinguished from all other siblings by higher self-esteem. This was more true of Jewish male only children than of Protestant or Catholic male only children. As Morris Rosenberg had also shown that high grades and high self-esteem are associated, there is support here for a connection between the only children's college attendance, their grades, their self-esteem, and, we may perhaps conjecture, their need for achievement. The self-esteem of the firstborn (as distinguished from the only child) was not superior to that of the later born in Rosenberg's study. No comments are possible about the last born, who are not clearly distinguished from his various categories of later born.

In sum, there are indications that only children (and perhaps first-born) are driven to school grades, to college, and to eminence by a need to achieve. The question then arises as to the peculiar nature of the parent-child relationships for only children and firstborn that produces this need for achievement. Are parents of only children more achieving themselves, so that they model achievement? Do they pay more attention to achievement in their offspring, and so reward achievement? Do the offspring become more achieving in order to please their parents? Although these alternatives sound simple enough, the answers to be supplied in the next chapters suggest that the matter is considerably more complicated.

affiliation and conformity

Recent literature suggests that it may not be possible to explain the achieving characteristics of only children and firstborn without prior reference to their affiliation and conformity. For this reason an examination of the parent-child characteristics that presumably lead to achievement must be delayed until the assertion that only children and firstborn are also more affiliative and conforming has been confirmed.

The studies of affiliation, as well as the current resurgence of ordinal position studies as a viable activity for psychologists, are directly attributable to *The Psychology of Affiliation*, by Stanley Schachter (1959). In this work Schachter pursued some aspects of social comparison theory, a theory which suggested that one basis for people's gregariousness was a need for self-evaluation (Festinger, 1954; Latane, 1966). Schachter's procedure was to induce varying degrees of anxiety about impending experimental events in a group of subjects. He wanted to know whether in these circumstances the subjects would prefer to be with others or to be by themselves while waiting apprehensively for the experiments to begin. Because the subsequent results seem to have a great deal to do with the particular nature of this situation, it is well to recapitulate his fear-inducing instructions at this point; although on the surface, they read like a parody from *Mad* magazine.

> In the high anxiety condition, the subjects, all college girls, strangers to one another, entered the room to find facing them a gentleman of serious mien, horn-rimmed glasses, dressed in a white laboratory coat, stethoscope dribbling out of his pocket, behind him an array of formidable electrical junk. After a few preliminaries, the experimenter began: "Allow me to introduce myself, I am Dr. Zilstein of the Medical School's Department of Neurology and Psychiatry. I have asked you all to come today in order to serve as subjects in an experiment concerned with the effects of electrical shock. . . . What we will ask each of you to do is very simple. We would like to give each of you a series of electric shocks. Now, I feel I must be completely honest with you and tell you exactly what you are in for. These shocks will hurt; they will be painful. As you can guess, if, in research of this sort, we're to learn anything at all that will help humanity, it is necessary that our shocks be intense. . . . Again, I do want to be honest with you and tell you that these shocks will be quite painful but, of course, will do no permanent damage [Schachter, 1959, p. 13].

The degrees of anxiety felt by the subjects in these circumstances were assessed on a 1–5 point self-rating scale (from "I dislike the idea of being shocked very much" to "I enjoy the idea very much"). Subjects then indicated whether they wanted to spend the time waiting for the experiment to begin, by themselves, with others, or whether they didn't care. Schachter's findings, which are of importance for the study of ordinal position, were that under conditions of high fear (there was also a low-inducing fear condition in which the experiments were not represented as being harmful), firstborn females showed a stronger desire to wait with others than did later-born females. Asked about a series of ascending electric shocks, only children and firstborn females were less willing to continue than were later born. When highly anxious firstborn and highly anxious later born were compared, the firstborn still showed the greater desire to be with others. There was the suggestion that while only children and firstborn were similar in their desire to affiliate, only children were, in general, less anxious than firstborn. The size of the subject's family made no difference, though the anxiety-affiliation relationship decreased in strength for each subsequent ordinal position. Subsequently, the subjects were told of the subterfuge in this experiment, given an account of the reasons for it, and, of course, no one actually received any electric shock.

"Speculating" on the origins of these differences, Schachter opined that in early childhood firstborn had perhaps been taught to associate the reduction of pain with the presence of others. The inexperienced mother rushing to the baby's side at every whimper, ready to interpret every burp as a death rattle, quickly conditioned the youngster to expect social solace in the presence of pain or fear. As the youngsters grew older and increased the magnitude of their own instrumental responses, they in turn sought out other persons when in states of pain or fear.

By contrast the later born, relatively neglected by the now much busier and wiser mother (who could tell a burp from a death rattle), learned that in states of pain or anxiety, tension reduction would have to come about mainly as a result of their own efforts. Although Schachter (1959) was most tentative about this developmental speculation—". . . if this sort of proposition amounts to anything [and] we do not seriously wish to defend these particular arguments" (p. 43)—it has become the vaguely held functional assumption in most subsequent work. Schachter continues: "Without examining further the merits of these arguments, this general line of thought does lead to the expectation that under anxiety provoking conditions first born and only children will manifest stronger affiliative needs than later born" (p. 43).

Intrigued by these relationships, Schachter reexamined various other sets of data in the literature and came up with some surprising findings. In a survey of chronic male alcoholics, it was found that they were more likely to be later born. Again, examining data on psychotherapy, it was found that firstborn were more inclined to go into psychotherapy and less inclined to drop out of it (Wolkon, 1968). They were also perhaps more inclined to suicide (Lester, 1966). Finally, in data on effective fighter pilots, it was found that the aces—that is, those with the more awards for destroying enemy planes—were more likely to be later born. Helmreich (1968) has subsequently found that later born also make better aquanauts and Nisbett (1968) that they participate more in dangerous sports. Linking these phenomena to the earlier results on affiliation, Schachter (1959) suggested that firstborn were more likely to seek the help of others when in states of anxiety (thus, resorting to psychotherapy), and the later born were more likely to handle their anxieties in isolation either effectively (as fighter pilots) or ineffectively (as alcoholics). Before we proceed further, it is important to demonstrate that the matter is worthy of scientific attention. In Table 6.1 we present the results of a number of experimental investigations of the thesis that in situations the same as, or somewhat similar to, those described above by Schachter, there are differential reactions by birth order. It should be stressed that a considerable simplification is involved in the results as presented in the table. That is, if the investigator set out to test some variety of the anxiety-affiliation hypothesis, and if he got birth-order results favoring the firstborn for one or more of his experimental conditions (even though he may have failed on some of his other conditions), the study is reported as a confirmation. Still, even with these qualifications and the more detailed ones to be presented later, the amount of confirmation for

TABLE 6.1 Tests of the Anxiety–Affiliation Hypothesis

Investigators	Year	Subjects	Results
Schachter	1959	Females	Confirmed 1B
Wrightsman	1960	Females	Confirmed 1B
Sarnoff & Zimbardo	1961	Males	Confirmed 1B
Gerard & Rabbie	1961	Females	Confirmed 1B females
		Males	Not confirmed 1B males
Radloff	1961	Females	Confirmed 1B
Weller, L.	1962	Females	Not confirmed
Zimbardo & Formica	1963	Males	Confirmed 1B
Ring, Lapinski, &			
Braginsky	1965	Females	Confirmed 1B
Eisenman	1965	Females	Confirmed 1B
Singer & Shockley	1966	Females	Not confirmed 1B
Gordon, B. F.	1966	Females and	Confirmed 1B females
		males	and males
Darley & Aronson	1966	Females	Confirmed 1B
Latane	1966	Females	Confirmed 1B
Miller & Zimbardo	1966	Females	Not confirmed 1B
Helmreich & Collins	1967	Males	Confirmed 1B
Becker, G.	1967	Females	Confirmed
		Males	

the thesis that firstborn affiliate more rapidly than the non-firstborn is quite impressive (see Table 6.2).

A second large group of experimental studies similar to the other and also derived from social comparison theory have focused on the thesis that in conditions of uncertainty, the firstborn will be more likely to conform to the opinions of others. Schachter (1959) has presented it in this way: "When together with others ... in a situation some aspects of which require evaluation . . . early born are more likely than later born to rely on others in evaluating their own opinions and emotional state" (p. 32). Both the *anxiety-affiliation* thesis and this *uncertainty-conformity thesis* were supposed to derive from the same special relationship of firstborns to their parents. In Table 6.3 we summarize the results from the latter thesis. Once again, notwithstanding the variety of only loosely

TABLE 6.2 Confirmation of the Affiliation Thesis

	Confirmed	Not Confirmed
Females	10	3
Males	5	1
Total	15	4

TABLE 6.3 Tests of the Uncertainty-Conformity Hypothesis

Investigators	Year	Subject	Results
Erlich	1959	Males	Confirmed 1B
Staples & Walters	1961	Females	Confirmed 1B
Dittes	1961	Males	Confirmed 1B
Sampson	1962	Males	Confirmed 1B
Becker & Carroll	1962	Males	Confirmed 1B
Becker, Lerner, & Carroll	1962	Males	Confirmed 1B
Moore	1964	Males	Not Confirmed 1B
Amoroso & Arrowood	1965	Females	Confirmed 1B
Carrigan & Julian	1966	Males	Confirmed 1B
		Females	Not confirmed
Darley	1966	Females	Not confirmed
Becker, Lerner, & Carroll	1966	Males	Confirmed 1B
Sampson & Hancock	1967	Males	Confirmed 1B
		Females	Not confirmed
Bragg & Allen	1967	Males	Confirmed 1B
		Females	Not confirmed
Hamilton	1967	Males	Not confirmed
		Females	Not confirmed

analogous experimental conditions, the degree of confirmation is striking, although the results suggest confirmation only for males, rather than for females (see Table 6.4). Actually the data show that firstborn males, together with firstborn and second-born females, are relatively conforming, whereas second-born males are not. Thus the ordinal difference holds for males. But this should not obscure the fact that all females are relatively conforming.

The consistency in the two sets of results, as shown in Tables 6.2 and 6.4, which give us a grand total of 26 positive experimental outcomes to 11 negative outcomes makes Schachter's major hypothesis, both with respect to the experiments and to development, worthy of much more detailed consideration.

TABLE 6.4 Confirmation of the Uncertainty-Conformity Thesis

	Confirmed	Not Confirmed
Females	2	5
Males	9	2
Total	11	7

ANALYSIS OF THE EXPERIMENTS

The intention of the following analysis is to discover, if possible, what it is that has been confirmed, and, alternatively, why it is that there are as many nonconfirmations as are reported above, particularly for the females in the conformity studies.

To begin with, only 2 of the 36 studies reported above control adequately for ordinal position (Sampson & Hancock, 1967; Bragg & Allen, 1967), an inadequacy which, it will be argued later, makes a critical difference in the outcome of these studies of affiliation and conformity. Each set of studies has also a number of other shortcomings and these will be dealt with in turn.

Anxiety-Affiliation Studies

First, with respect to the affiliation studies: What are the instigating conditions in these experiments? What leads the firstborn subjects to express a desire for greater affiliation than the later born? As described at the beginning of this chapter, the instigating condition was one of anxiety (fear or threat) about an impending shock. So we may begin with the question of whether firstborns are initially more fearful or anxious. Do they arrive in the laboratory with a more fearful set? Results from self-report measures of general anxiety at the college level are inconclusive (Weller, L., 1962; Moore, 1964; Jacoby, 1968). But measures taken within the experimental situation, whether direct by galvanic skin responses (Gerard & Rabbie, 1961) or by self-ratings in response to anticipated electric shock or by indication of willingness to participate in further shock experiments, fairly unequivocally show the firstborn to be more unwilling, "anxious," "upset," and to perform at lower levels under stress (Schachter, 1959; Wrightsman, 1960; Eisenman, 1965, 1966; Yaryan & Festinger, 1961; Helmreich & Collins, 1966; Helmreich, 1968; Nisbett & Schachter, 1966; Carman, 1899). Furthermore, such childhood reports as there are, fairly universally quote the firstborn as showing more anxiety than the later born (Sears, Maccoby, & Levin, 1957; MacFarlane, Allen, & Honzik, 1954; McArthur, 1956; Rosenberg & Sutton-Smith, 1964b). In the Berkeley Guidance data on children between the ages of six and sixteen years, the firstborn, in particular M1M and F1M are consistently more anxious throughout. Put together with the earlier cited paranatal results (Weller & Bell, 1965; Weiss, 1967a), we may perhaps conclude that the firstborn arrive at the experiment in a condition which predisposes them to anxiety.

Do they also arrive with a stronger affiliative set? The self-report measures that have been taken of the affiliation motive have been quite contradictory. In one study, firstborn—males and females in Cincinnati —showed higher need affiliation as measured by a thematic procedure

(Dember, 1963); in another, on the contrary, second-born males at Tufts University showed more affiliation than firstborn, using a similar device; although when a verbal report device was used, the firstborn showed more interest in affiliation (Conners, 1963). The fact that the character of the self-report device can change the scoring, and the lack of control for ordinal categories prevents any safe conclusion being drawn from these studies. Sampson and Hancock's (1967) results for the specific ordinal positions and the two only children give us a rank order of need affiliation scores as follows, with the highest scorers being mentioned first: F1M; F; FF2; F1F; MF2; M1M; FM2; M1F; MM2; M. Females scored more highly, as we might expect, but there were no significant differences. If this particular order is fairly stable, however, and we shall advance reasons later on to suggest that it is, then it can be seen that the alternations of firstborn and second born throughout would mean that a slightly higher numerical weighting of subjects in any one of the ordinal categories could throw the results of a firstborn versus second-born comparison in either direction. Perhaps in the Tufts University example above there were more later-born males with older sisters, and in the Cincinnati sample more first-born males with younger brothers! We shall conclude simply with the statement that while it is possible that some ordinal categories have higher need affiliation than others, a simple distinction between firstborn and non-firstborn oversimplifies the actual ordinal differences. The experimental outcomes then may be based in part on a greater readiness for anxiety or affiliation arousal in some subjects as a result of preexisting sibling status differences. These phenomena are not controlled by the experiments, and, indeed, despite the points made above, we do not know much about them.

What is it then within the experiments that seems to lead to sibling status differences? Fear of electric shock has been the central variable, and in most cases, the presentation of the threat of an impending electric shock has differentiated between the birth orders. If the intensity of threat is made even greater, however, it is apparently possible to eradicate this birth-order difference. Miller and Zimbardo (1965) failed to obtain such differences when it appeared they were going to take blood samples, electroencephalographic readings and, as well, intentionally left blood-stained clothes scattered about the experimental room. When fear is great enough, it induces avoidance responses in all subjects.

A number of experiments have been able to produce affiliation behavior with a variety of other conditions. Sarnoff and Zimbardo (1961) produced the same reaction in males (at Yale) by implying that they would have to engage in regressive behavior, such as that of sucking nipples and baby bottles, as a part of the impending experiment. Such regressive anxiety as well as direct fear, they found, had the same birth-order effect. In addition, there has been another group of experiments in which the same result has been induced by creating a condition of ambiguity where the subjects lacked information (Radloff, 1961), had reason to doubt their opinions (Gordon, B. F., 1966), felt unsure of their own abilities (Hakmiller, 1969), and lacked self-esteem (Zimbardo & Formica, 1963). So we have *fear* and *uncertainty* as two alternative instigating conditions in these various experiments upon affiliation.

Second, given these instigating conditions, what is it that the subjects did? Schachter was able to show that under conditions of anxiety, subjects would affiliate. He was not able to demonstrate whether they came together mainly for direct anxiety reduction through the comforting presence of others (emotional comparison) of a similar state, or whether they came together mainly for cognitive self-evaluation (opinion comparison). Much of the subsequent research has been concerned with this emotional-opinion comparison dilemma.

Zimbardo and Formica (1963), who favored the emotional comparison viewpoint, gave subjects in a Schachter-type situation the opportunity of going with others who were still in the process of completing the experiment or with those who had completed it. Highly anxious subjects, and highly anxious firstborns in particular, chose to be with those in the same emotional state as themselves. Ring, Lipinski, and Braginsky (1965) inserted a single firstborn or second born in a group with two other stooges who acted out various levels of emotionality. The firstborn more than the second born changed their self-rated emotionality in the direction they perceived in the accomplices. "We conclude that firstborns are more influencible than later borns and more likely to reduce discrepancies between themselves and others by moving towards them" (p. 12). An important finding was that in this experiment it did not matter in which direction the accomplices were placed (high or low excitement)— the firstborn saw themselves as similar to them. They also said they liked them and felt closer to them. Latane, Eckman, and Joy (1966) similarly found firstborn showing more preference than later born for those with whom they had had a stressful emotional experience.

In contrast, when offered as alternatives—those with a similar personality but a different experiment, or those of the same experiment but with a different personality—most subjects chose the same personality (Miller & Zimbardo, 1965), demonstrating, according to these investigators, that the subjects did not seek social comparison because of their experimentally provoked emotional uncertainties, but that they sought the presence of those who were most similar in general (Becker, 1967). It seems then that the type of social comparison that will be made depends very much on who is available, so that the problem is not so much one of a choice between emotional and opinion comparison as between the degrees of similarity in both these dimensions that one can obtain with the available target persons.

It would be misleading, however, to oversimplify the matter. While some form of similar other is more often sought by firstborn in situations of anxiety or uncertainty, our knowledge of the reasons involved is imprecise. In the general social comparison literature, for example, it has been shown that subjects tend to choose similars when they want to compare themselves with others on positive characteristics (Wheeler, 1966); but when negative characteristics are involved, they prefer to compare themselves with inferiors for the purposes of self-enhancement (Hakmiller, 1966); or they may prefer superiors for modeling purposes when relatively unfamiliar characteristics are involved (Thornton & Arrowood, 1966).

The delicacy of the relationship is further indicated by a variety of other isolated findings. Thus, Helmreich and Collins (1967) found that when subjects were in a state of fear, the firstborn preference for being with peers extended only to waiting for the experiment to occur, but not for performing in it, when, like the later born, they preferred the company of someone in authority. Furthermore, in a number of studies with a slight alteration in conditions, experimenters have been able to produce a preference in firstborn for being alone, rather than with others (Suedfeld, 1964; Darley & Aronson, 1966). So, while we can say that, given the instigating conditions of anxiety or uncertainty, firstborn will more often choose to be with similar others, there are many things about their preference that are not clear.

Perhaps more fundamental as a criticism is the neglect throughout this series of studies of any consideration of the "representative" character of the experimental situation (Brunswik, 1956). These experiments are miniature social situations, and one of the most important significances to be derived depends upon the logical parallel to be drawn between them and other representative social situations. For example, none of the 37 experiments entertained any controls for the experimenter as a variable, and yet the corpus of birth-order studies strongly suggests that there are birth-order differences in responses to adult power figures. Lack of control for the experimenter as a variable (Rosenthal, 1964) introduces the strong possibility that the consistent results across the varying conditions above may have been brought about by the subjects' special and differential attitudes to the experimenter rather than to the explicit experimental conditions. The contradictory birth-order evidence on *volunteering*, for example (firstborn more often volunteer for psychology experiments), seems to show that such responses most often arise when the power figure makes a personal appeal to the subjects. When appeals are made on a mass basis, birth-order differences in willingness to volunteer for experiments seem less likely to occur. At least, this is one interpretation that can be placed upon the positive results (Capra & Dittes, 1962; Eisenman, 1965; Suedfeld, 1964; Varela, 1964) as compared with the negative results (Brock & Becker, 1965; Schultz, 1967; Wagner, 1968; Ward, 1964; Wolf, A., 1966; Weiss, Wolf, & Wistley, 1963). Again, in the conformity studies, to be dealt with below, controls are never exercised for the fact that accomplices in the Asch situation are peers. Yet there is strong sociometric evidence that the birth orders already differ in *peer acceptance* as measured by popularity (Alexander, 1967; Finneran, 1958; Patterson & Zeigler, 1941; Sells & Roff, 1963; Schachter, 1964). If peers already evoke differential responses from the birth orders, their use in conformity experiments is not a neutral condition. On both these counts, then, in these 37 studies experimenters and peers treated as neutral conditions may well have contributed to the variance in ways which affected the results but were outside the control of the investigators.

In sum, the instigating conditions in these experiments may presuppose a birth-order differential readiness for anxiety and affiliation, may presuppose a miniature social situation affording a special relationship to a power figure (the experimenter) or to peers, and clearly do provide for conditions of threat and uncertainty within the experiment itself.

Uncertainty-Conformity Studies

Fortunately, if we turn from the anxiety-affiliation thesis to the uncertainty-conformity hypothesis, there is a considerable amount of additional information that assists the process of interpretation. It is not that the conformity experiments are any more precise in outcome than the affiliation experiments, but from the combination of the two we begin to get a general picture of what may be going on. Once again, this review stays with the experimental studies. Reference to non-experimental studies shows the usual amount of contradiction. "Conformity" is not a general condition of firstborn which will automatically manifest itself under any circumstances, as the contradictory literature on "acquiescent response set" (Weiss, R. L., 1966), and the "social desirability response set" (Newfield, 1966) makes clear.

First, some description of the peculiarities of the conformity experiments is in order. Like the affiliation experiments, no control is generally exercised for the specific ordinal positions nor for the possible influence of the experimenter. In addition, in the conformity studies, most of which involve peer accomplices, all experimenters proceed as if such accomplices are (apart from their experimental perfidy) a neutral influence.

As some of the conformity studies have been done with children, we shall begin with those. Carrigan and Julian (1966) used 96 sixth-grade children in Buffalo as their subjects. They compared boys with girls, firstborn and only children with later born. Subjects were asked to make judgments as to which of several stories most adequately suited a picture projected on the screen. They were informed throughout which of the stories "last year's class from this school picked as the most fitting story." The measure of conformity was the extent to which they made the same choices as the ones which were supposed to have been the choices of last year's subjects. In another condition of the experiment, one group was also administered a sociometric test just prior to the judgment task. They were asked to make a list of their best friends in this group. It was expected that having to think about whom "you" were choosing and who might be choosing "you" would increase the social anxiety of the group members. It was expected in particular that such social anxiety would have a greater effect upon only children and first-borns, because of Schachter's work showing relationships between anxiety and affiliation.

The results indicated that firstborn were indeed more conforming under both conditions, but particularly under the anxiety condition after the administration of the sociometric test. But most of the difference between the conditions was produced by the males. Later-born boys were both less conforming and less affected by the difference between the conditions of the experiment. Although the mean scores for females were higher than those of males, there was no significant differentiation between the female birth orders. Both firstborn and later-born girls responded like firstborn boys. This outcome perhaps implies that at age 12, "conforming" in this study was a sufficiently higher order response for *all* girls that birth-order differences did not show.

A similar result was achieved by Becker and Carroll (1962), using 48 boys (age unspecified) on a Chicago playground. Again the firstborn showed more willingness to yield to what they thought was majority opinion than did the later born.

More typically, conformity situations have involved the following type of circumstances (Becker, Lerner, & Carroll, 1966). Students (first-born and only children versus later born) are brought to the experimental situation and they are asked to say which of three lines is "closest in length to the line on this card (the experimenter indicates a standard card). Don't try to tell me how long the line is, but just which of these lines most closely matches this line" (Becker et al., 1964, p. 321). The experimenter has child accomplices in the group, however, and these accomplices volunteer their incorrect opinion before the subject gives his own opinion. So there is group pressure to make a judgment that coincides with that of the accomplices, even if it does not seem correct to the subject. Typically the firstborn yields to the erroneous opinions of the accomplices in this condition more than do the later born. There are a number of other studies of conformity where achievement variables have also been involved, but these are better discussed later when relationships between the two variables are the focus.

Unfortunately, there has not as yet been any test of whether those of a given birth-order position who affiliate are also those who conform; though there are some general indications that dependency and conformity are related in groups at large (Darley, 1966). Nevertheless, having made these qualifications, we can possibly derive certain conclusions both from the anxiety-affiliation and the uncertainty-conformity series. Given the fact that all these experiments are themselves miniature social situations in which authoritative people (the investigators) are doing important things (experiments), in which one can make use of or must rely upon peers, it appears that the conditions of anxiety and/or uncertainty induced have a greater effect upon firstborn. They do seem to be more aroused by these social situations. They more often seek out or attempt to join with what others are feeling or doing. We would emphasize, however, that these experiments may have been fruitful because of the conditions they did not control in creating their social situation as well as the conditions they did control.

The important question for the moment, however, is whether the results of these experiments, and the interpretations we have put upon them, justify Schachter's developmental thesis. Could the mother's trepidating concern with her firstborn have produced these outcomes? Have her uncertainty and anxiety been adopted by the firstborn? Have the firstborn in turn learned to depend upon the facilitation and help of others as a result of this initial attention? Is their concern for social comparison and their exercise of social consensus an adult derivative of such early events? A way of checking on this possibility is to make an examination of the developmental literature to see to what extent the observations that have been made on mothers, infants, and children actually bear out these developmental assumptions. At the same time this will permit an examination of the questions with which the previous chapter ended, concerning the developmental origins of achievement.

developmental transformations of dependency and achievement

There are not actually any developmental studies of affiliation, conformity, and achievement. What we have are a few scattered studies on parallel topics at different chronological age levels, and from these we must draw inferences with respect to Schachter's thesis. Still, it makes more sense to draw inferences about childhood primarily from studies made on children, and secondarily from studies made on adults.

The most relevant study to deal with infant behavior was one carried out by Cushna (1966) at a pediatric clinic in Iowa. His subjects were middle-class children from 16 to 19 months of age. They were observed immediately after an immunization shot in the clinic and observed also in their own homes. The investigators were interested in the children's reaction to the shot, and in their reaction to the mother's going out of the room in the home setting. They observed whether the child cried, whether he sought out his mother when she left the room, and so on. The results were fairly clear: at home the firstborn were more upset by the mother's leaving them, and in the clinic setting the firstborn boys were more upset than the firstborn girls. In both clinic and home settings the children were asked to carry out a variety of verbal and motor tasks, such as throwing a ball, climbing on a chair, responding to a request to jump, and so forth. The mothers were asked to assess the child's performance level on a variety of items drawn from the Vineland test of social maturity. In both cases the mothers' expectations of firstborn far exceeded their expectations for later born. The overassessment was most pronounced with firstborn boys. The mothers' expectations for independence training, for example, for firstborn were two standard deviations above the mean for that age group! And in fact, the firstborn did better at the motor performances than did the later born. In addition the mothers were more active in prompting and helping the firstborn, though in different ways for the two sexes. They were more supportive, strategic, alluring, and more cautious in directing their boys. But (particularly in the home) they were more demanding, exacting, and intrusive in telling their firstborn girls how a solution is reached, a finding similar to that of Rothbart (1967) with 5-year-olds and quoted in the section on opposite sex effects in Chapter 3. In another study of infants, Gewirtz and Gewirtz (1965) found that mothers interacted *twice* as much with only children as with the youngest.

Bayley's results with intelligence tests are of a consistent character (1965), as are the recent results of Solomons and Solomons (1964) with motor performances. As long as it is remembered that we are actually talking about tests of motor and sensory performance and that these have not been shown to have a strong predictive relationship to intelligence as it is measured later, then Bayley's results can be treated like those of Cushna, as indices of the way the babies are responding to additional social stimulation at that age level. Bayley's report on 1,409 infants ranging in age from 1 to 15 months showed small but consistent findings in favor of the firstborn in 11 out of the 15 months. They were significantly higher on her "mental" scale in 4 of the 15 months and on the "motor" scale in 5 of the 15 months. The later born were not significantly higher on any of the months.

Bayley (1965) says: "If we assume that there are differences, they could be accounted for if we postulate a greater amount of parent-child interaction with first borns, thus stimulating them to more rapid development" (pp. 391–392). But she goes on to say: "There is, however, no indication of any cumulative effect of such stimulations; the differences between first born and later born for both ages are small, and they tend to be pretty well distributed over the entire age range" (p. 392). Thus it appears that the initial stimulation affects the infant scores but has no subsequent affect on intelligence tests.

In these initial results, then, we do find, as Schachter suggested, the mothers more concerned with firstborn's performances, but also we find the firstborn performing at higher levels and being made more anxious by separation from the mother. One gets the impression that while the mother's concern for the firstborn might lead the firstborn to refer back to the mother more often, what they get from her is further expectation of higher level performance. If anxieties are assuaged by this mother attention, it appears likely to be only one part of the relationship. Again, one is impressed not so much by the mother's anxious inexperience as by the adultlike demands she is making on her firstborn as compared with her later born. The higher level of expectation by the mothers of the boys' achievement and perhaps of good behavior from girls is worth restating. It could be that the mother's expectations of both are consistent with sex role expectations because the girls are simply more malleable, and the boys more resistant to their expectations (Moss, 1967). The girls can be ordered about, but the boys must be lured. A psychoanalytic formulation would, of course, explain the same phenomena in terms of the mothers' rivalries and loves.

In addition to Cushna (1966) and Gewirtz and Gewirtz (1965), the only other observational study of what parents actually do (there are a number on what they say they do) with infants was carried out by Lasko (1954) in Ohio. The mothers were observed in their homes and rated on their different treatment of their firstborn and second born. There were 46 pairs of siblings who were the firstborn and second born in their families, and who were observed in a longitudinal study at repeated intervals. This meant that the mother's treatment of the siblings could be compared at the same chronological age for each sibling. Comparisons were made between the ages of one and nine years. There is a striking confirmation of Cushna's observations insofar as the firstborn are the subject of much greater verbal and intellectual accelerational attempts in the observations made during the first two years. The parents prompt more and satisfy their curiosity more. After the age of two, this is no longer the case. Lasko (1954) says: "By the age of three or four, however, the home no longer revolves around him and starting from a much more favored position in the beginning, he is less warmly treated than is his younger sibling at a similar age" (p. 115). In these and subsequent years, furthermore, the firstborn continues to receive more attempts at acceleration and more "disciplinary friction" from the mother. The age correlation data most strongly suggest that the major change for firstborn over the years is the very substantial decrease in the available interaction with their mothers. Understandably enough, the mothers have less time for them. Non-firstborn received more consistent treatment throughout, and Lasko also concludes that as mothers deal with later children—second, third, and so on—they are somewhat warmer and more protective toward them, but also more arbitrary and strict.

There are a number of points here for further discussion. First, Lasko did not find mothers more anxious with their firstborn than with their second born. But like Cushna, she did find them more demanding. Rather than thinking of the parents as always running to the child because of their own inexperience (each burp a death rattle), it might be a simpler thesis to interpret the adults as treating the infants as much like adults as they are able. The parents of firstborn have been used to dealing with other adults; they have seldom had experience with infants. If they simply generalize these customary responses, then they will expect a very high level of performance from their infants and then believe they perceive more problems in their behavior (Shrader & Leventhal, 1968). They may not expect so much from their later born, partly because they have less time, but partly also because by the time the second born arrives, the parents have been educated by their firstborn and have scaled down their expectations. A good illustration of the way in which parents can be educated by their infants, although in a reverse direction, is provided by Harlow's experiments with those monkey mothers who had not themselves benefited from real monkey mothering. They were instead reared on a bottle mechanically manipulated in a wire cage. These monkey mothers with this nonprimate rearing experience so ignored their own firstborn that several of the infants were mutilated and some killed. But the monkey infants were so persistent in seeking feeding opportunities that their mothers forcibly learned a great deal about mothering by association with them. The mothers were, in consequence, much more effective mothers with their own later-born infants. In this example the newborn monkeys had educated their mothers in an upward direction. It is not hard to believe that firstborn humans educate their parents in a downward direction, forcing them to revise unrealistic expectations derived from associating too long with mature people. Of course, it can also follow that the mother's high level of demand makes the firstborn more anxious about pleasing her (even if she herself is not more anxious with the firstborn than with later born). Some have argued, for example, that the firstborn may be much conflicted by the reduction of attention after the first two years (Sears, Maccoby, & Levin, 1957).

Another point to note is that although Lasko and Cushna demonstrate particular additional attention to firstborn in early years, Lasko also demonstrates the continuance of a special type of interaction with firstborn in subsequent years. Unfortunately, her parent-child rating scales are not designed to be sensitive to the later childhood years, but the existence of more "disciplinary" friction between parents and firstborn suggests higher expectations are still being held out for them.

There are a number of reports by parents of the way in which they say they differentially treated their children by ordinal position, but we wish to give these studies a secondary status. It might seem unfair to treat mothers as more subjective observers of themselves and their own children than are such investigators as Cushna and Lasko, but there are, unfortunately, good reasons for putting the mothers on trial and demanding that they first prove their case. Longitudinal studies of mothers' reports of their children's behavior seem to indicate that they are not a particularly reliable source of evidence (Radke-Yarrow, 1963), at least as time goes by, for they appear to distort their earlier accounts in terms of their current attitudes. Still, it has to be admitted that in the current evidences, mothers' reports seem to be consistent with the data already given above. Thus, most investigators have found that the mothers report that they expected more of their firstborn. Gewirtz (1948) found them more concerned with their firstborn. Sears, Maccoby, and Levin (1957), in their very comprehensive interview study with 379 mothers from the New England area, however, reported that the mothers said they were more indulgent of later children only in large-size families, but not in two-child families, a discrepancy which is unresolved at this point, and perhaps has something to do with the higher levels of achievement aspiration in small-size families than in large ones (Rosen, 1961).

This section on parent-infant relations may be summarized by saying that apparently both parents and firstborn are more concerned with each other than are parents and non-firstborn. By concern, we must at this point mean higher expectations by the parents (Lasko, 1954; Cushna, 1966; Stout, 1960; Gewirtz, 1948); higher performances by the children (Cushna, 1966); more anxiety about being separated from the parents by the children (Cushna, 1966); and more sensitivity to pain by the boys (Cushna, 1966). Later we shall cite findings of Rothbart (1967) and of Hilton (1967) that parents also interfere more and are more inconsistent with firstborn. But in addition, we call attention to the suggestions of a continuing "special" relationship between parents and firstborn denoted by greater "disciplinary friction" (Lasko, 1954), "more chores" and "more conscience" (Sears, Maccoby, & Levin, 1957).

When we move beyond infancy, with some exceptions (Hilton, 1967; Rothbart, 1967) there tend not to be studies of what parents do to the children (as this varies with ordinal position). Instead, there are studies of what children of different ordinal positions are like. If one adopts an infant-deterministic thesis—that is, that what parents do to infants molds their personality permanently—then of course the parents' role has already been explained, the child's radarscope has been set, and the parents' subsequent relationship to the child is, therefore, more or less irrelevant. But there are a few studies which suggest that parental influences continue to count, and these are taken up in turn.

But first we shall deal with the other objective studies of children's characteristics as they vary by ordinal position. In 1948 Gewirtz reported an observational study of preschool children at the Iowa Child Welfare Research Station. His hypothesis accords nicely with the observations of Cushna (1966) and Lasko (1954), whose studies nevertheless succeed his in time. The hypothesis was that as parents have more time for only children and early born than they have for later born, they will more often reduce their tensions and thus will become more reinforcing to them. This is not dissimilar from—though broader than—Schachter's (1959) notion that firstborn would find parents (and subsequently others) reinforcing because the parents consistently reduced their infant pain-derived tensions. Both Gewirtz and Schachter see the firstborn reaching out first for the parents and later for others because this reaching out has been rewarded by the parents. Lasko's and Cushna's studies, however, show us mainly the parents' coaxing and urging on the firstborn, which is a form of tension-enhancement, not reduction. Whether such increase of the child's level of aspiration is also accompanied by a decrease of his physical and other tensions through comforting actions, we do not actually know, though comforting actions might well be expected to accompany any such attempts to push the child ahead rapidly if they were to be successful.

In any case, on the basis of his thesis, Gewirtz (1948) predicted that only and firstborn would reach out toward others and that they would have a high succorant need, more often than later born, because this had become one of their most effective instrumental behaviors. From this point of view, such succorant behavior need not be connected with anxiety, as Schachter had postulated.

Gewirtz (1948) made observations on 45 children of ages 4 and 5 years at a nursery school. Over a period of three months, each child was observed some 16 times for 15 minutes each time. Gewirtz classified the observed behaviors into three groups. Positive attention-getting, which was taken to be a direct expression of succorance or help-seeking, was expected to be manifested mostly by firstborns and only children. Seeking reassurance and bodily contact with others was taken to be an indirect expression of succorance and expected to be manifested by later born who had not received so much reinforcement for their direct approaches to the parents. Finally, negative attention-getting was conceptualized as conflict expression of the need for succorance. Here, it was conjectured, the child's need for succorance, being strong but mainly rejected, led the child into a reverse and hostile expression, getting attention from others by rejecting them in turn.

The three categories of succorance were, indeed, found to be statistically independent of each other. The results were that, as predicted, only children showed succorant behavior more often than children with siblings. Firstborn with siblings, however, expressed their need more often by staying near to others and by seeking reassurance. They were also the least aggressive of all the groups. When we recall Lasko's observation that such firstborn received a sudden decrease in attention with the advent of the next siblings, these Gewirtz results appear to show that the firstborn are now (4–5 years) in a rather precarious position: most concerned with adult reassurance, but not willing to express antagonism in any way. The middle born siblings, when compared with the others, showed more negative attention-getting behavior, suggesting that they had had the least attention of all, and were being driven into rather hostile modes to get some minimal attention from adults. The youngest were more like the only children.

Gewirtz' study is particularly important because it introduces differences between the various ordinal positions, and is not just a dichotomous approach as is true of most birth-order studies. Again, it begins to highlight the particular incongruities in the development of the firstborn. The firstborn in Cushna's study (1966), mentioned above, were at that time mainly only children, although most of the mothers said they intended to have more children, and probabilistically most surely did, which means that firstborn, like only children, began by having higher expectations; but then, unlike only children, suffered a shift in parental rewards and were apparently moved themselves in the direction of more precarious expressions of succorance, that is, seeking reassurance but not expressing hostility. The middle born who had never been seduced into believing in parents as rewarding figures (we speculate), but like all infants and preschoolers needed things only adults could give, perhaps struck back less covertly by being a nuisance.

In any case, at this point we develop the propositions that only children (and younger born) will tend to show an uncontaminated expectation of help from others; they will, therefore, use affiliation as an instrumental behavior. But firstborn, having suffered a mild defeat for the same expectations, will seek more reassurance and comfort, so that they come to use affiliation as a consummatory behavior or end state (in itself). And again middle born will show an aggressiveness toward getting attention, rejecting affiliation as a means behavior.

Work similar to that of Gewirtz is reported by Haeberle (1958) and was carried out in 1958 with mildly disturbed New York City children three to six years in age in a therapeutic nursery school. Children were rated by observers for their dependency behavior (touching, attention-seeking, recognition-seeking). The mean scores for boys and girls follow consistently down the ordinal positions in the same way as Gewirtz' scores for succorance. Thus:

	Boys			Girls	
0	1B	LB	0	1B	LB
24.68	19.24	16.16	23.48	21.29	20.31

That girls had higher scores in general probably follows from the fact that mothers find that paying attention to girls leads to greater changes in their behavior, which in turn tends to keep the mothers paying more attention to girls than they do to boys. For example, when the baby cries and the mother consoles it, if it is a girl, it is more likely to stop crying; if it is a boy, it is more likely to keep right on crying. Infant girls are more rewarding to their mothers than infant boys (Moss, 1967). On the other hand, if we follow this same line of reasoning, what then is so rewarding about only boys? Why are they so dependent and succorant? Are we to believe that mothers of only boys are particularly persistent in rewarding them, despite their masculine recalcitrance? Later evidence which suggests that the only boy has marked identification with his mother makes this a worthwhile conjecture.

More recently Hilton (1967) has found that in an experimental situation where only children, firstborn, and later-born 4-year-olds worked on puzzles, the mothers of the only children and firstborn were significantly more interfering, extreme, and inconsistent in their behavior than were the mothers of the later born. This is consistent with Rothbart's very similar observational study (1967) with 5-year-olds in which the mothers put more direct pressure for achievement on their firstborn, were more intrusive into the activity of the firstborn (who were trying to solve puzzles), and yet were more likely to help the firstborn. In Hilton's study the mothers of the early born more often signaled to their children to begin work on the puzzles; the mothers of the later born were more likely to wait for the child to start by himself; again, the mothers of the early-born children continued to give more task-oriented suggestions, but they also made more overt demonstrations of love, especially if the child was doing well; if he was not doing well, they markedly decreased both their love and their verbal support. While various interpretations are possible, Hilton (1967) suggests:

> When the parent is inconsistent there is no stable guideline for internalizing the correct course of action. The child cannot predict outcomes on the basis of past performance, and must continue to ask for evaluation because the same behavior will elicit a varying response. The effect of excessive interference is to create standards that the child must fulfill. He does not set his own goals but rather achieves the ones set for him [p. 288].

Hilton goes on to say: "Interference and inconsistency both undermine the child's opportunities to develop reference points for internal evaluation" (1967, p. 288), which is to imply that these children are continually in the position of the subjects in Schachter's (1959) social comparison experiments, that is, reaching out for reassurance and information from others. Her type of argument makes social comparison theory once again particularly appropriate to firstborn. Unfortunately, there is no indication as to why the mothers should be so inconsistent. Perhaps their early high expectations followed by their forced reduction in attention, their use of the older sibling as a surrogate, as well as that sibling's continuing dependency—all combine to establish this vacillating pattern in the mothers. In a subsequent study (1968), Hilton has shown that parental interference does not work effectively unless the child is already dependent. A dependent child is comfortable with an interfering adult and works effectively as a result! But an independent child is only disoriented by this interference. Such a view leaves room for Schachter's more primary notion that the dependency is, in the first place, established by the additional attention and expectation (and presumably reinforcement) focused by mothers upon firstborn children.

The studies on early childhood therefore are loosely consistent with Schachter's developmental thesis. The parents may not relate to their firstborn in quite the way he suggested, but it does seem apparent that they do have a special type of relationship with them. To this point, therefore, there is fairly good evidence for believing that the consensus in the adult experimental results and the consensus in these studies of young children are connected. In the adult studies, the firstborn rely upon others and achieve more than the non-firstborn; and in these infant studies the firstborn also reach out for others and in most studies achieve more than do the non-firstborn. Inconsistencies over time (Lasko, 1954) and in immediate treatment (Rothbart, 1967; Hilton, 1967) may increase the importance of achievement as an instrumental behavior, and affiliation as a consummatory behavior for firstborn children.

CHILDHOOD

Even if a special parent relationship is postulated for parents and firstborn in preschool years, such a relationship may not continue subsequently. The evidence suggests, however, that it does. From this it can be argued that if ordinal position differences do persist in affiliation and achievement, this may well be a continuance of the early determinations dealt with above. First, we shall present some of the rather tenuous evidence of continuing special treatment and, second, further evidences of ordinal position differences in affiliation and achievement throughout childhood.

There are a number of studies all of which add up to the view that, in general, the firstborn continue throughout childhood to be the subject of special expectations on the part of the parents and to have a special relationship with them. Thus Stout (1960), using the Berkeley Guidance data (chiefly ratings based on interviews with parents), found that each parent was more directive of firstborn than later born. Parents tended to act jointly toward the firstborn but differentially toward the later born. Koch (1956a) found that firstborn at the age of 6 years spoke more clearly and articulated more adequately than second born, a difference that she put down to their more consistent modeling after and interaction with adults rather than siblings. She also found firstborn more concerned about parental alignments, relations, and favoritism. Bossard's transcripts of family table talk (1945) demonstrated that younger children "tend to be ignored . . . the family seems to adjust its age level to the older children, and to ignore the younger ones, especially if the age differential is not large. Questions on word meanings raised by the younger children are given less consideration, even in our most intelligent families" (p. 230). McArthur (1956) quoted a series of studies involving interviews, anecdotal observations, and content analyses of interviews, all of which also added up to a continuing "special" relationship of firstborn and parents. He began by citing the anecdotal and unpublished observations of Lantis concerning 134 children of Harvard students upon whom she collected life history data. Lantis summarized her impressions as follows: "The eldest child is adult-oriented. He is more likely to be serious, sensitive (that is, his feelings are hurt easily and he doesn't need much punishment), conscientious, 'good,' fond of books or fond of doing things with adults. . . . The second child is not so anxious to get adult approval; in this sense he is tougher" (McArthur, 1956, p. 48). In a subsequent study reported in the same paper (McArthur, 1956) in which the parents of these children checked the elements of these descriptions which were most characteristic of each birth order, a significant number concurred in attributing the characteristics in much the same way as Lantis had done. Later, in a still further interview with the parents of these parents (grandparents of the original subjects) which was also scored for the same descriptive items, first children were described by both generations as serious and adult-oriented and second children were described by both as nonstudious, cheerful, placid, and easygoing. The trait "shy" was characteristic of firstborn and "friendly" of second born. A majority of McArthur's parents reported being more relaxed in their handling of the later born, but there were no significant associations between specific child training procedures and birth order differences. He suggested that despite important historical changes in child training over recent years, "the resulting personalities in each generation seem to have been the same, the first child was adult-centered, the second child was peer-oriented" (1955, p. 52).

To test whether firstborn children of grade school age did in fact perceive themselves as interacting to a greater extent with parent figures, Houston used the Bene-Anthony Family Relations Test (Sutton-Smith, Rosenberg, & Houston, 1968). Subjects were 40 New Zealand boys of ages 5 through 11 years of differing sibling positions but matched for age, intelligence, and socioeconomic status. All were members of two-child families (M1M, M1F, MF2 or MM2). An adapted version of the test

(1957) consisting of 40 items—ten of positive outgoing feelings from the child to his family; ten negative outgoing; ten positive incoming; ten negative incoming—was used (see Table 7.1). Each item was printed on a card, read to the child, and given to him. First, however, from an array of 20 line-drawn figures (four men, four women, five boys, five girls, a toddler, and a baby) the child selected figures representing his family. An extra box, Mr. Nobody, was added by the investigator. The items read to the child in random order were mailed by him in the mailbox behind each of the family figures he had chosen.

The results indicated, as predicted, that the firstborn boys placed more messages in the parent boxes than did the second-born boys. Specifically the firstborn recorded more positive intake from the father $(p < .05)$, and more negative intake from the mother $(p < .05)$. The second born for their part showed more interactions with their older siblings than vice versa. They received a more negative intake $(p < .01)$, and gave more negative responses $(p < .01)$ to the siblings. For both sibling orders where there was the greatest interaction, therefore, there was also the greatest conflict. This study, like a number of others, again points to the involvement of the parents and the firstborn.

Let us turn to our second question: Does the differential parental treatment lead to continuing differences in affiliation and achievement? The answers available indicate that indeed ordinal position differences continue. But they seem to imply also that the relationships between these variables may change. We shall begin the discussion with an experiment by Gilmore and Zigler (1964), the complexities of which serve to heighten the contradictions that occur if one stays with the view that the dependency responses of infancy will simply persist in a similar form throughout development. In a study with five- to eight-year-olds, Gilmore and Zigler (1964) proposed that as firstborn "have been continually satiated on social reinforcers early in their lives, they will be less effortful in a situation where they can take these for granted and more effortful in a situation where they cannot" (p. 193). From the studies of Gewirtz (1948) and Haeberle (1958), above, this is a proposition which we might expect to be more true of only children, firstborn, and non-firstborn, in that order. Unfortunately, Gilmore and Zigler tested it with only 20 firstborn and late-born boys and girls. More important, where Gewirtz argued that the preexistence of an abundance of social reinforcement would lead to more succorance behavior in the firstborn three- and four-year-olds, Gilmore and Zigler were saying that the same condition would lead to less effortful behavior in the five- to eight-year-old firstborn.

Gilmore and Zigler used the "marble hole game," which is nomenclatural doublethink for a monotonous and repetitive procedure, in which the subject sorted 300 marbles of two different colors into two

TABLE 7.1 Items in Adaptation of the Family Relations Test

Positive Outgoing Feelings:

1. This person is always very nice.
2. This person is nice to play with.
3. This person is kind hearted.
4. This person is jolly.
5. This person often helps the others.
6. This person is lots of fun.
7. This person deserves a present.
8. I love this person very much.
9. I would like to keep this person always near me.
10. I would like to sit on this person's knee.

Positive Incoming Feelings:

11. This person likes to play with me.
12. This person is very kind to me.
13. This person makes me feel very happy.
14. This person likes to help me.
15. This person thinks I am a nice boy.
16. This person smiles at me.
17. This person often wants to be with me.
18. This person always listens to what I say.
19. This person likes me very much.
20. This person likes to give me things.

Negative Outgoing Feelings:

21. This person spoils other people's fun.
22. This person is bad tempered.
23. This person is not very patient.
24. This person is sometimes too fussy.
25. This person sometimes makes me feel very angry.
26. This person sometimes grumbles too much.
27. Sometimes I don't like this person.
28. Sometimes I hate this person.
29. Sometimes I would like to spank/smack/hit this person.
30. This person is a nuisance.

Negative Incoming Feelings:

31. This person hits me.
32. This person teases me.
33. This person scolds me (tells me off).
34. This person won't play with me when I feel like it.
35. This person won't help me when I am in trouble.
36. This person is too busy to have time for me.
37. This person is always complaining about me.
38. This person makes me feel sad.
39. This person gets cross with me.
40. This person makes me feel foolish.

SOURCE: B. Sutton-Smith, B. G. Rosenberg, & S. Houston, *Sibling Perception of Parental Models.* Paper presented at Eastern Psychological Association, Washington, D.C., April 20, 1968.

different holes in a boxtop for as long a period of time as he wished to continue. The experimenter said: "Now in this game you can put as many marbles in the hole as you want to. You tell me when you want to stop." Under one condition of playing the game, half the firstborn and later born were given no encouragement while they proceeded. In the other condition they were encouraged: "You really know how to play this game." "Fine." "Good." The results were that the firstborn played for a longer period of time when they received no encouragement than when they were encouraged. Later born played longer when they were given encouragement than when they were not given any. The nonencouragement condition stimulated the firstborn to greater effort; the encouragement condition stimulated the later born to greater effort. Gilmore and Zigler concluded that "firstborns evidenced less need for social reinforcement than did later borns when such reinforcers were readily available" (1964, p. 199). Later born who accelerated their performances when the reinforcers were available were said to have done so because they were relatively deprived in earlier years.

These results should perhaps be taken seriously because there is partially corroborative data in Koch's complex study of the 8 two-child family positions at age six. She had teachers rate her subjects on various traits including their relationships to the teachers (1955b). One of her traits, "response to sympathy and approval from adults," may be equated with the encouragement condition in the Gilmore and Zigler study. If we follow Gilmore and Zigler's line of reasoning, then under normal conditions the second born would show more of such a response, being hungrier for any available reinforcements than were the firstborn. The rank order of the levels of response for the various two-child ordinal positions at the different age spacings are indicated in Table 7.2, an adaptation from a table by Koch (1955b, p. 20).

TABLE 7.2 Average Ratings on Response to Approval

Ordinal Categories	Age Differences between Siblings		
	7–24 months	25–48 months	49–72 months
MM2	5	3	2
FM2	8	4	8
MF2	3	1	7
FF2	1	2	3
M1M	4	8	5
M1F	6	6	4
F1M	2	7	1
F1F	7	5	6

SOURCE: Adapted from Koch, H. L. Some personality correlates of sex, sibling position, and sex of sibling among five and six year old children, *Genetic Psychological Monographs*, 1955b, **52**, 3–50.

It will be noticed in this table that only under the intermediate age spacing does the expected difference between firstborn and second born hold up. In that spacing, second born are always more responsive after receiving the rewards of sympathy and approval from adults. This is an important substantiation of the expectation that second borns will be more susceptible to immediate rewards because this two- to four-year-age gap is the one in which a definite contrast will occur between the mother's handling of the two children. This contrast between the siblings is muted in both the nearer and further spacings. It is noticeable that the effect seems to work most consistently across the various age spacings when the second born is one of the same sex as the firstborn (FF2 and MM2). Here, sex of sibling is not a complicating variable as it seems to be in the other second-born cases. The virtue of these supplementary findings from Koch is that although they can give only partial support, or partial refutation (whichever way you choose to look at it) to the Gilmore and Zigler material presented earlier, they do introduce examples of the type of sibling position complexities with which research in this area must actually deal. Still having conceded that Gilmore and Zigler have some partial corroboration in Koch, how then are their findings to be reconciled with those of Gewirtz and Haeberle? Are the differences due to the different age levels? Do younger firstborn children, because of their immaturity, express their need for reinforcement more directly? Or is the nursery school, with the inevitable competitiveness for the teacher's approval and help we especially associate with four- and five-year-olds, really more like the nonreward condition in Gilmore and Zigler's experiment? If both situations are indeed comparably frustrating, then we might say that at both age levels the firstborn make a greater effort—the younger in Gewirtz and Haeberle by reaching out for succorance, and the older in Gilmore and Zigler by working longer when not rewarded. It might make sense to equate both the nursery school and the marble game in the nonencouraging condition because they are both slightly frustrating situations, and make sense also to see direct appeal (four- to five-year-olds) and appeal through effort (five- to eight-year-olds) as differential age responses to a similar stimulus complex.

Whatever the case, with the adolescent studies it is possible to argue that the relationships between these variables has taken yet a further shift. In the Becker, Lerner, and Carroll experiment (1966), quoted in the previous chapter, it was shown that the firstborn boys conformed more in an Asch perceptual judgment situation, agreeing that the size of the line was the same as had been stated by the accomplices. In one of the conditions in the 1966 experiment, however, there was no resulting difference between the ordinal positions in yielding to accomplices. Subjects were told: "I'll give you a chance to win some money here. You will get a bonus of $.05 for each accurate answer you give" (p. 321). The experimenters had already established that this was regarded as a not very significant amount of money by these boys. Second-born responses here were like their responses in the earlier control conditions. They did not take much notice of the accomplices. But now the firstborn also ignored the accomplices. Why? Did the desire for $.05 make them rely on themselves more? Did the money now make their own achievement more salient for them than the opinion of the accomplices? Unfortunately, Becker, Lerner, and Carroll also shifted their instructions slightly. They now stated the task not as a perceptual judgment of matching lines, but as a matter of accuracy of judgment. It is possible that this shift in instruction shifted the meaning of the experiment to a concern with a psychological function (judgment rather than perception), about which the first might feel more personally involved. There is some evidence that firstborn are less competent at perceptual judgments (Koch, 1954; Steward, 1967).

In two other experimental conditions, the firstborn showed less conforming behavior than the later born: first, when the reward offered was 25 cents, and second, when the judgments had to be made after the cards were taken away and the subjects had to rely on their own memory in the face of the incorrect statements of the accomplices.

Partial resolution of some of these differences has been provided in a recent study by Rhine (1968). He hypothesized that in low achievement conditions the firstborn would be more conforming (after Schachter, 1959), but in high achievement conditions where they could profit by independence of judgment they would be nonconforming (Sampson & Hancock, 1967). In an experimental situation where the subject had to judge the number of clicks in a series that he had heard after hearing the judgments of others, the firstborn followed the others when they were not rewarded for accuracy, but stayed with their own judgment when they were. That is, under nonachievement conditions they were conforming, but under achievement conditions they were nonconforming.

While there are too many differences among these various studies to be certain that the varying results are not produced by varying instructions, the psychological functions involved, or the social roles of accomplices and experimenters (Carrigan & Julian, 1966; Helmreich & Collins, 1967; Erlich, 1958), there does seem to be some merit in the view that with increasing age, the relationships of affiliation and achievement shift. Perhaps affiliation or dependency has a primary role with young children, so that achievement is a means to that end; whereas, with the passage of time, achievement becomes a more important motive in its own right. It stands to reason that when one's own career is paramount (as in adolescence) rather than one's relationship to parents, the need for achievement would take on a more autonomous role in human behavior. At the same time the particular success of the firstborn and only born in becoming eminent might well imply that they may ultimately use their affiliative skills in service of their achievement. Possibly the relationship between the two variables is reversed during development, so that dependency as a form of "servility" in childhood gives way to dependency as a form of "sensitivity" in adulthood (Alexander, 1967).

In Chapters 5, 6, and 7 it has been established that, in general, the firstborn are more achieving, affiliative, and conforming. There seems, as well, considerable evidence for the view that this is brought about by their special and continuing relationship with their parents, who expect more from them and treat them in an inconsistent manner. The firstborns' continuing need for reassurance and guidance appears to evoke contradictory behavior from the parents. The child seeks help but performs well. The parent gives help but is critically expectant of an even higher level of performance. Each member in this uneven synchrony is calibrated to the other. There appears to be evidence also that the relationship between the variables of affiliation and achievement changes as the subjects get older with affiliation being more important in the earlier years and achievement in the later. As will be shown, however, the description offered here is more true of some firstborn and second-born contrasts than of others.

CHAPTER 8

hierarchical and egalitarian interactions

The previous chapters have dealt with several classes of variables (sex role, power, affiliation, and achievement) which have been covered extensively within the literature of ordinal position. There has been less attention in that literature to several other classes of variables (conservatism, popularity, and empathy) whose status is less certain. These variables are considered in this chapter although in a most tentative way. The previous chapters have examined the data largely in terms of siblings modeling after each other; of siblings counteracting each other's influence; and of parents shaping firstborns' behavior in certain systematic ways. These types of influence have been considered within the contexts of sibling-sibling interactions and parent-child interactions. Little attention has been given to the mutual relationships of sibling with sibling and of sibling with parent.

The present chapter takes up this larger social structural problem, approaching the firstborn, for example, both in terms of their relationship to parents and in terms of their relationship to younger siblings. It would seem doubtful that sibling differences can be understood without studying within the one conceptual grasp the types of learning that occur (modeling, operant, counteractive), the content of those learnings (affiliation, conformity), and the interaction patterns (child-child, parent-child), which provide the social context for both of these. It is the intent within this chapter, therefore, to use some of the newer types of data mentioned above (empathy, conservatism, popularity) in an attempt to trace tentatively some of the larger patterns of interaction within which sibling status may take on its significance.

HIERARCHICAL INTERACTIONS

The data and speculations on affiliation and achievement presented in the previous several chapters appear to support the view that parental actions have been an important shaping influence (the operant paradigm) in accounting for firstborn distinctiveness. The mother expecting higher levels of achievement, paying much attention to her firstborn as an infant, inconsistently rewarding and demanding, over time shapes up a child who reaches out for her and for others who provide social guidance and partial reinforcement. Most of the data presented can be considered to be consistent with this account—although, as pointed out earlier, physiological birth differences in firstborn may give them a more important instigating role with their inconsistent mothers than is apparent at present. We know little about the infants' contribution to this mother-child synchrony, and as the mothers' inconsistencies are also something of a puzzle, it is clear that our knowledge of the total relationship is, at best, quite partial.

In the previous chapters it did not seem necessary to consider the alternative explanation that the parents themselves provided the examples of achieving and affiliative behavior for the child to copy. And yet given the usual power of parents, it might be assumed that some such modeling took place. If so, it appears that it would provide a supplementary rather than a primary explanation. There is not at present any substantial evidence to suggest that all mothers of firstborn are themselves affiliative or achieving, and, therefore, models for these characteristics. This is true, even though middle-class mothers model these characteristics more than lower-class mothers (Rhine, 1969). Still, before tracing the total pattern of interactions of which the firstborn are a part, it is necessary to examine in some detail some of these other potential relationships to their parents.

There are other phenomena to be considered which would imply that while the shaping action of the mother is an important early influence in explaining firstborn behavior, it is not a sufficient one. For example, in the material already supplied, the developmental account of the changes in the relationships of affiliation and anxiety would seem to imply that as the children got older, other contingencies entered into this patterning of influence. Thus the connotations of achievement appeared to change their nature with changing chronological age. An achieving response instrumental in pleasing parents at one age may be relevant later mainly as an autonomous means of gaining success in the larger world. New rewards and new values may alter the relationships of the variables earlier patterned by the actions of the mother and the infant needs of the children.

Again, there are a number of studies in the literature which seem to suggest that firstborn children model after their parents along lines other than those already suggested. It is often said, for example, that firstborn children *identify* with their parents, by which it is usually meant that they do more than simply copy parent behavior; they also want *to be* like their parents. They both copy the parent and have a motive *to be* the parent. These are interesting distinctions and sound like common sense. But at this point in the study of sibling status it is still an important preliminary to discover empirically in what ways firstborn behavior does actually parallel that of the parents.

It has been argued that if firstborn model after parents while they are young and immature, they might be expected to be more "primitive" in this copying and to end up, in consequence, more rigid, conservative, and conventional in subsequent behavior. Whether or not such a modeling or identification process takes place, there is scattered evidence that firstborn are more conservative (Altus, 1967; Hall & Barger, 1964; Eisenman, 1965; Kammeyer, 1966). Others have argued that the firstborn will have a harsher superego, be of greater conscience or be more highly socialized through their internalization of adult values at an earlier age. Again there is some scattered support (Harris, 1964; Harris & Howard, 1966; MacDonald, 1967, 1969a, 1969b; Palmer, 1966; Kayton & Borge, 1967; Sears, Maccoby, & Levin, 1957).

Another prediction is that if the firstborn are more like adults they will tend to process information in a relatively adult way. They will show a preference for verbal and conceptual over nonverbal and perceptual modes of information processing. Contrarily, with less verbal contact and information from parents and siblings the second born will follow the principle of "matched-dependent" behavior (Miller & Dollard, 1941), which involves empathy based on the forms or actions of the model (Harris, 1964). Firstborn, Harris suggests, rely on the principle of "copying," which involves empathy based on verbal feedback from the model and leads to understanding of the thoughts and judgments behind the external forms. Non-firstborn should be more successful in empathizing when their attention is directed to external forms of behavior, while firstborn should be relatively more successful in empathizing when their attention is directed to the reasons which lie behind such forms, says Harris. As a test of these hypotheses the present investigators chose the Stick Figures Test and the As-If Test (Sarbin & Hardyck, 1955). The Stick Figures Test was designed to investigate the influence of postural behaviors on role perception. The test consists of 43 line drawings in various postures presented along with a multiple choice set of items for selecting the action or feeling conveyed by the posture. Subjects' responses can then be judged against the modal responses of the standardization group.

The As-If Test is a measure of verbal inference. It can be described as follows. The questions are asked: "In what way would your life be different if you had been (first) (second) born? The test was scored for the centrality of inference of the responses using a 1–9 point scale. This scale permits a judgment as to whether the subject is paying attention to more trivial aspects of another role or to more central aspects. Examples of scoring categories were as follows: (1) describes a real difference in self-concept, for example, "mature earlier," "more independent," "less anxious"; (5) describes a difference in role behavior, for example, "less attention from parent," "care for younger sibling," "date earlier"; (9) describes a difference in social or physical environment, for example, "gone to a larger school," "less material things," "more outside activities." Each subject's answers were scored independently by two raters with interrater reliability of .91.

Both tests were administered to 60 male and female college subjects, 15 of each in the 8 two-child family positions. Four separate simple analyses of variance were conducted. For the male groups on the As-If measure, ordinal position was a significant source of variance ($F = 3.67$, $df = 3/56$, $p < .05$). (See Table 8.1). Although in two of the analyses the between groups effect ($F = 2.02$, $df = 3/56$, $p < .10$, and $F = 2.10$, $df = 3/56$, $p < .10$) did not reach acceptable levels of significance, multiple comparisons by t test were undertaken.

The Stick Figures Test scores differentiated between the second-born boy (MM2) and his firstborn brother (M1M), with the former being superior ($t = 2.24$, $df = 28$, $p < .05$). However, the relationship between the boy with the younger sister (M1F) and the boy with the older sister (FM2) was not significant. For girls, those with brothers (FM) were superior to girls with sisters (FF) ($t = 2.17$, $df = 58$, $p < .05$).

In the case of the As-If Test the predictions were supported for males, but not for females. As can be seen, male firstborn were superior to male second born ($t = 2.65$, $df = 58$, $p < .05$)—the relationship obtaining most strongly for the firstborn boy with a younger brother (M1M). Thus M1M > MM2 ($t = 3.25$, $df = 28$, $p < .01$); also M1M > FM2 ($t = 3.47$, $df = 28$, $p < .01$). In addition, though not predicted, M1M > M1F ($t = 2.28$, $df = 28$, $p < .05$). Comparisons for females did not attain significance, but the tendency was for girls with sisters to be

TABLE 8.1 Summary of Mean Scores on Two Empathy Measures by Group

Group		Stick Figures Inference	As-If Test
Male with younger brother	(M1M)	10.40	3.60
Male with younger sister	(M1F)	11.33	4.80
Male with older brother	(MM2)	12.53	5.47
Male with older sister	(FM2)	10.67	5.27
F ratio		2.10	3.67
p		$<.10$	$<.05$
Female with younger sister	(F1F)	12.60	4.07
Female with younger brother	(F1M)	14.33	4.00
Female with older sister	(FF2)	13.40	3.73
Female with older brother	(MF2)	13.93	4.53
F ratio		2.02	
p		$<.10$	ns

NOTE: On the Stick Figures Test, a high score is indicative of greater empathy. On the As-If Test, a low score indicates greater empathy.
SOURCE: Rosenberg, B. G., Sutton-Smith, B., & Griffiths, J. Sibling differences in empathic style. *Perceptual and Motor Skills*, 1965, **21**, 811–814.

superior on the As-If Test to girls with brothers, the reverse of the finding with the Stick Figures Test.

These results, then, support the notion that methods of inference vary by ordinal position, but only for males. Male firstborn use the predicted adult method (verbal inference) in a superior way. Male second born use the predicted later-born method (perceptual inference) in a superior way. There are a number of other studies in the literature which also suggest a difference of this sort, namely, Koch's discovery (1954) that second born are superior at tests of perceptual speed, Steward's (1967), and Altus' (1965) findings that later born are superior at the embedded figures test which requires careful attention to perceptual details, and Altus' argument that second born prefer art and music courses, because the learning is through demonstration rather than by words (1967a). Contrarily, emphasis on the verbal superiority of firstborn (Cushna, 1968) and their greater skill in fantasy may be regarded as of similar import (Singer, 1966). The differences between males and females are puzzling. But there is a parallel with the age spacing and cognitive data cited in Chapter 3 of the present book in which males were also differentiated by ordinal position but females by the sex of the sibling. We offer the speculation that ordinal position differences in cognition are greater for males because in general the male is more enduringly affected in his cognitive development by his emotional tie to the mother. There is longitudinal evidence for such greater enduring effects of mothers' treatments on males (Yarrow, 1964). In females, cognitive performances seem to be less tied into their mother-child relationships and more responsive to the social situation in which the females find themselves. It is as if the female is capable of generalizing her dependency on the mother to subsequent persons (even to brothers), but the male remains tied to the original relationship. If this interpretation is correct, then the present data support the notion that the firstborn males are modeling after their mothers, in respect to the cognitive competences manifested on the As-If Test.

In an attempt to gain some insight into female modeling after adults the present investigators studied 80 females (20 in each ordinal position) with the Gough Adjective Check List which included a 300-item pool of adjectives. Subjects were asked to check those items they felt were self-descriptive and those they believed to be descriptive of adults. All parent and self scales were converted to equivalent scores, and the difference scores between the parents and self by sibling group were compared. This comparison was made in terms of eight scales previously derived from the Gough Adjective Check List. Table 8.2 gives the levels of response by each ordinal category to parent and self-description.

Further analysis of the significant interaction ($F = 3.23$, $p < .05$) by t tests indicates that the firstborn perceived themselves as most identified with their parents and the second born as least identified. These data nicely complement the Bene-Anthony data given in Chapter 7. In addition, the scoring of the particular need scales shown in Table 8.2

TABLE 8.2 Responses to Adjective Check List (Gough)

(Ns = 20, each)

	Achievement	Aggression	Autonomy	Dominance	Favoritism	Nurturance	Order	Self-control
SD	48.6	46.6	47.6	48.0	51.4	51.3	49.2	52.2
F1F								
PD	52.1	46.7	43.2	53.1	54.2	54.5	57.7	57.3
SD	48.6	51.4	51.3	48.8	50.6	48.5	51.7	49.6
F1M								
PD	57.4	53.9	53.1	60.4	59.0	50.9	54.3	52.5
SD	49.6	52.0	54.1	52.7	50.0	46.9	48.0	46.5
FF2								
PD	55.1	53.7	48.3	55.8	56.2	52.3	57.4	56.6
SD	49.3	51.3	49.5	48.5	47.6	48.1	51.6	47.2
MF2								
PD	53.3	52.5	48.2	55.1	50.9	51.7	54.2	51.9

NOTE: SD = self description; PD = parent description.

indicates that the firstborn girls were higher on nurturance, order, and self-control, whereas the second born were higher on aggression and autonomy. As these descriptions match our stereotypes about the differences between adults and children, they provide some further confirmation of the notion that firstborn girls model after adults with respect to expressive traits.

In sum, in their conventionality, conscience, superiority on verbal measures of inference (males), and self-descriptions as orderly, nurturant, and self-controlling (females), these firstborn show considerable evidence of being more similar to adults than do later born and thus provide some evidence for the claim that parents not only shape firstborn behavior (Chapters 5, 6, and 7), but also provide models for some of it. The discrepancies introduced between males and females, however, make it very clear that there is much yet to be learned about the way in which this modeling differentially affects the sexes of the children (as well as the birth orders) with respect to different types of variables. MacDonald (1967, 1969) has presented the thesis that although firstborn in general model more closely after parents, each sex models its appropriate sex role traits so that they often end up being quite different from each other. On these grounds, firstborn females should exhibit responses which subserve sex-role typical dependency, and firstborn males should exhibit responses which subserve sex-role typical independence. Unfortunately, the cross-sex data presented in Chapter 3, Table 3.11 would seem not to support this argument.

Yet another explanation (beyond shaping and modeling) which might be required to explain some firstborn behavior refers to the experiences they have as the dominant siblings. Whether they are dominant by virtue of their own power, or because they have been delegated the dominant role by their parents, they do have a long experience of superiority over younger siblings. Bossard and Boll, in their study (1955) of personality roles in the large family, for example, claimed that such families usually led to the development of a number of specialized roles. They said:

> Every family identifies at least one of its siblings as the responsible type, the one that is looked up to, the one that assumes direction or supervision of the other sibling, or renders service to him. The word responsible is the one used most frequently in referring to these siblings, but in some cases such words as dutiful, bossy, drudge, leader, helpful, martinet, and policeman are also used. These seem to identify chiefly the way in which this position of responsibility was exercised. . . . [p. 102].

The most clearly and frequently identified of all these responsible siblings is the oldest or older daughter who becomes in varying degrees a second mother to the younger children" (1955, p. 102). Bossard and Boll then proceed to discuss the "popular, sociable, well-liked sibling" (usually the second born), "the socially ambitious type," "the studious ones," and "the isolates." While Bossard and Boll's role analyses involve broad descriptive characterizations and are essayistic rather than quantitative, there is a consistency between their emphasis on the relative "responsibility," "surrogation," and so forth, of the firstborn, the similar claims of many other writers, and the traditional position of firstborns in primogeniture. The prediction should be possible, therefore, that when adult occupations can be categorized in terms of their surrogate character, firstborn, particularly females, should be found more often occupying roles for which they have thus been prepared by their childhood experiences of nurturance, responsibility, and surrogation.

Some tests of these predictions are available. In our own study (1964) of ordinal position and occupational preference, firstborn college girls showed a significantly greater preference for the surrogate role of teaching, a finding replicated by Fischer, Wells, and Cohen (1968). Hall and Barger (1964) have shown that firstborn have a preference for controlling and organizing the behavior of others, and there are in addition some data from not very reliable personality inventories demonstrating that firstborn characterize themselves as more dominant (Bender, 1928; Abernathy, 1940).

We have, then, the situation that the firstborn are more anxious, affiliative, and achieving toward adults, more like adults in being conservative, of high conscience, and preferring verbal means of role inference; but at the same time they are powerful and domineering in their relationship to their subordinates. These various descriptions suggest that we can tentatively formulate the first child as one who is involved in a network of hierarchical relationships.

Stotland (1968) has conceptualized these differences by suggesting that the point of view of the firstborn is that within the family the parents are the most important (as nurturing, controlling, and of high power). The other siblings come along later and act interference, but do not substantially alter the firstborns' perception of a hierarchically arranged social system. The firstborns' perception of the family system, therefore, is one in which power is controlled by those of high status, who are the nurturers, and upon whom one is dependent. The firstborn both models and enacts a hierarchical social system. The later born, by contrast, comes into the family, which already has two types of more powerful people—the parents and the siblings—the latter being somewhat similar to the second born himself. His concept of the social system, therefore, is a more variable one, partly hierarchical, partly egalitarian, from which the prediction may be drawn that while firstborn will find it easier to relate to others in a hierarchical way (alternatively dependent or dominant), later borns will find it easier to relate in an egalitarian manner. From this view it should follow that firstborn would be more interested in hierarchizing any social situation, adopting a dependent role if in an inferior status and a dominant role if in a superior status. There is tangential evidence that may perhaps be read in support of a proposition of this sort. Firstborn have been shown in the experimental literature to be more concerned with discovering what the group norm or consensus is (Dittes, 1961; Singer & Lamb, 1966), or to want the guidance of a superior (Storer, 1961). They are more likely to choose popular people sociometrically (Schachter, 1964; Weinstein, 1967). In a study of siblings' best-friend playmate choices, it was found that the firstborn were more likely to say that their best friend was someone who was more powerful than themselves (Sutton-Smith, 1966). All of these studies may be taken as indications of a firstborn seeking for authority in ambiguous social situations; seeking for a hierarchization of a social situation which would possibly allow them to be dependent upon the established leaders or norms. On the other hand, when in positions of authority, firstborn seem more likely to exercise it. When they know they do represent the group consensus, they make a greater attempt to influence others (Gordon, 1966) and to expel those who are deviant from the group (Arrowood & Amoroso, 1965). As teachers, firstborn (males) have been found to be more critical, disapproving, and hostile (Solomon, D., 1965), but at the same time to also be more nurturant in a reward situation, giving out more reinforcements to others (Weiss, 1966). This is a contradiction interestingly reminiscent of the inconsistencies in the mothers of firstborn children in the Hilton (1967) and Rothbart (1967) studies.

What these various speculations and fragments of data add up to perhaps is recognition that they may be all-reconcilable within a social structural framework which takes adequate account of the varying role sectors in which each sibling type is involved. The firstborns' case is the most complex. There are the mutual influences of being shaped by adults, modeling after adults, and yet exercising both surrogate and actual power over less powerful siblings. Perhaps, as Stotland (1968) implies, none of these forms of learning makes much sense unless they are conceptualized as mediators for the content of the hierarchical social interactions within which the firstborn find themselves. Perhaps for the student of sibling status it is not enough to know that shaping or modeling or counteraction are involved. Also required are reference to the prevailing types of interaction and the specific types of response within these. But this is a topic which we shall take up again when some of the materials on the later born have been introduced.

EGALITARIAN INTERACTIONS

Strictly speaking, later-born interactions are of many sorts—not solely egalitarian. The term *egalitarian* is used here for its contrast rather than comprehensive value. The smaller concern which the later born have for hierarchical interactions may be said to be responsible for their greater popularity with their peers.

The evidence is not completely clear-cut, but it has tended to favor later over firstborn. Sells and Roff (1963) used a sociometric device in which children chose from their class groups those they liked most and those they liked least. Some 1,013 children in two Texas school districts in Grades 3, 4, 5, and 6 were used. The mean scores for the different birth order groups are shown in Table 8.3. The line between the third and fourth ranks was a natural and statistically significant division between two clusters along an acceptance-rejection dimension according to Sells and Roff. Here we do find the second born more popular than the firstborn, but the youngest and only children are even more popular— paralleling their respective positions on college entrance data given in Chapter 5.

The college level materials, by and large, also seem to support the notion that the non-firstborns are more popular. They have been rated "better mixers" (Patterson & Zeigler, 1941; Storer, 1961); more sociable and well-liked (Bossard & Boll, 1956a); popular (Schachter, 1964; Finneran, 1958); more open in interpersonal relationships (Dimond & Munz, 1968). Recently, however, Alexander (1967) has challenged most of this evidence, showing in a very large sample of high school males ($N = 1410$) that firstborn are more popular, and demonstrating with considerable cogency that the popularity of the second born at the college level is due to their being a more highly selected group—both with respect to socioeconomic status and popularity. It is much less customary for second born to go to college, Alexander argues, and when they do, they tend to be a "special" group, with motivation and characteristics similar to those of firstborn. Once again, more evidence is clearly necessary. But it is not inconceivable that the outgoing overt characteristics of later born actually do make them more popular in the childhood years, but that the more "internalizing" (MacFarlane et al., 1954) and sensitive

TABLE 8.3 Interactions of Siblings

Birth Order Group	Mean Scores
1. Youngest child	5.19
2. Only child	5.13
3. Second of two	5.09
4. Second of more than two	4.98
5. Oldest child	4.91
6. Middle child	4.87

SOURCE: S. B. Sells and N. Roff, "Peer Acceptance-Rejection and Birth Order," *American Psychologist*, 1963, 18, 355.

(Koch, 1955b) firstborn become popular as these become the more relevant characteristics of adolescent behavior. Perhaps as relationships between affiliation and achievement change, the firstborn become more skilful at using their interpersonal sensitivities. It is perhaps not worth speculating further on these differences in popularity, however, until there has been more adequate study with the specific ordinal positions, although we might expect MM2 and FF2 to be particularly popular second born in childhood because of their sex-role typical characteristics, modeled after their same sex older siblings.

Besides attempting to explain the later borns' aggression on the basis of their reaction to the firstborns' power, and their popularity from the fact that they model their roles from other children as well as from adults, we might attempt also to explain such evidences as there are of their interpersonal insight by the same experience of learning from other children. That is, we can seek to use the interactions of the later born with their elders to explain both their aggression, their popularity and their empathy.

Stotland (1968) has argued for example, that later born are accustomed to more egalitarian relationships than are firstborn. They use their older siblings as models of what they themselves can do (not as the elder must use the parents as models of what they should do), and are, therefore, more directly interested in those whom they perceive as similar to themselves. In several experiments Stotland and co-workers (1962, 1963) have demonstrated that if the later born perceive themselves as similar to the experimentally provided model, they are more likely to act the way they believe that he acted. Firstborn are not similarly affected. The models were generally said to have carried out a series of routine tasks (underlining letters) similar or dissimilar to those done by the subjects and to have done them well or poorly. Stotland maintains that the experiments were different from those cited in the chapter on affiliation and conformity in that other persons were not available for emotional or information self-clarification. There was no opportunity for feedback. The subject could not go to them to find out how he was doing. He was simply told whether or not he was performing the same tasks as they had and whether they (the models) were good or bad. If he was told he was performing the same tasks and believed the model to be good, then the later born worked more accurately and evaluated himself more positively than did the firstborn (Stotland & Cottrell, 1962).

In another experiment of a similar sort, later born were shown to identify more emotionally with the fate of the model than did the firstborn (Stotland & Dunn, 1963), and in yet another experiment, to identify with the model if they could interact with him, but not if he was not the locus of interaction (Stotland & Dunn, 1962). Stotland summarizes these various experiments by suggesting that later born empathize more with those whom they perceive as similar to themselves and with whom they can interact, which is, to Stotland, a replication of their developmental experience with their older siblings.

To confirm the notion that some emotional empathic process is involved, Stotland and Walsh (1963) have repeated experiments similar to those above, but with the subjects this time wired for physiological indices of emotion, namely, vasoconstriction (changes in heart amplitude) and palmar sweat. The outcomes vary, but in general, later born showed more empathy (as thus measured) as long as they perceived themselves as similar to the model who was the target for their observations. Results were somewhat different for later-born males and females. It appeared as if this empathic response was more easily induced in the females, and that the males required some preliminary "warming up" experiences to get them involved. When they were asked to imagine themselves in the other's position beforehand, the ordinal differences emerged.

While these are a most interesting group of researches, the many inconclusive findings and the marginal nature of the positive findings require a tentative response. In a study of our own in which we used conditions similar to those used in the Stotland and Dunn (1962) experiment, but varied the target model so that in one condition it was a peer and in another an adult, we did not gain any positive results for the hypothesis (Griffiths, 1966). Still, if on the other hand we assume that Stotland's scattering of marginal findings is meaningful and that later born do more readily empathize with similars, there are other materials in the literature which appear to be supportive of such a point of view. In particular, we would point to the factor analysis of college male sibling attitudes carried out by Hall and Barger (1964), part of which has already been referred to. One of the key factors for later born was "gregariousness," an activity which had the connotation of liking people for their own sake, and liking activity for its own sake. Unlike the first-born, they did not seek such gregariousness for purposes of leadership and control and not presumably (Schachter, 1959) as a response to stress. The gregarious activity seemed to be a more general system of responses for the later born, at least insofar as it accounted for a much larger part of the variance in their responses. Storer (1961) reports a similar conclusion in a contrast between M1M and MM2. One might argue that such attitudes are the type of outcome of sibling experience that Stotland would predict. Later born, using other siblings as models for their own impulses, would continue to show an interest and satisfaction in peer-prompted activity. Firstborn, modeling after their elders, would be concomitantly more interested in leadership and control, or in the instrumental use of others.

Pursuing this same line of argument, it might follow that since later born have more models (siblings as well as parents), they would have a great variety of social responses or roles at their command. They have had to react to these different persons, all of whom are important to them in very different ways, so that very different types of reactions would be a part of their own key social repertoires. This might mean that a later born would be more socially flexible or socially diffuse than a firstborn. The flexibility would occur when a positive use was made of the larger repertoire, and the diffuseness when it was inappropriate to the social requirements. We might argue contrariwise that the firstborn would be relatively more stable when socially appropriate and rigid when not. We have noticed above that firstborn do more hungrily seek the group consensus and more arbitrarily wield it when in power, perhaps supporting this notion that they are more likely to vary along a stability-rigidity dimension than a flexibility-diffuse one. Still, this is all very speculative. On behalf of the notion, Hall and Barger (1964) found one factor distinctive for later born which they denoted as a "flexibility" in the attitudinal structure of the younger born. They say: "The description of the younger siblings should be one emphasizing individuality and idiosyncrasy rather than conformity. Their entire factor structure is characterized by a quality of diffuseness described as versatility and flexibility, with a concomitant de-emphasis on seriousness of purpose and the need to control others" (1964, p. 66). Of a supporting nature also are a number of other studies which purport to show that later born (males) are more tolerant as compared with firstborn males, who are more critical and disapproving (Solomon, 1965). Somewhat more tangentially related are the studies indicating that aesthetic preferences (Eisenman, 1965) and performance (Eisenman, 1968) are more unique and creative among the later born, and alternatively that they take a longer time making up their minds about their major at college (Altus, 1967a).

Following this line of thinking, the present investigators were led to argue that if the later born do, in fact, play more roles, and are, in fact, more role diffuse, then they should, in general, be more interested in drama and acting. Henry (1965) had shown that actors, as compared with controls, tend to suffer from psychological states similar to those described by Erikson as "identity diffusion" (1950). Whether the later born might prefer to act roles because (a) they had a greater variety of role taking reinforced; (b) because they had less often received reinforcement for one stable surrogate role (as perhaps the first had); (c) because they were still seeking roles, never having been adequately reinforced for any of their previous roles, or (d) because they were, in consequence of these factors, in a state of conflict over the roles they usually adopted, is not certain.

Following our own methodological canon that only specified sibling groups should be dealt with in ordinal research, we investigated the relative proclivity for, and competence at, drama on the part of first-born girls with younger sisters in two-child families (F1F) and the younger girls from the same type of family (FF2) (Sutton-Smith, 1966c; Sutton-Smith & Rosenberg, 1966a). Two procedures were followed.

First, a group of sophomore college girls were asked to list the number of plays in which they had participated while at high school and college. There were 38 F1F's and 32 FF2's. The mean number of plays listed by the firstborn was 2.50 and by the second born 5.53 ($t = 2.65$; $p < .01$). This study was done in 1963. It should be noted here that Leventhal (1966) found that second-born boys (MM2) when compared with firstborn boys (M1M) also showed a greater interest in "theatre."

Second, in the spring of 1965, F1F's and FF2's were asked to play opposite each other in a role-playing scene. They were to be two sisters arguing over their future vacation spot. Concurrently and unknown to these actors, they were observed by other F1F's and FF2's through a one-way screen. The latter Ss made judgments as to which player in each pair had the best lines and which player had the best expressions. The actors' sibling positions were not known to the judges. The study involved 6 pairs of actors (F1F's and FF2's) and 4 pairs of judges (F1F's and FF2's). On speaking their lines, 18 judgments favored F1F, and 30 favored FF2 ($X^2 = 3.00$; $p < .10$); on expressions, again 18 favored F1F, and 30 favored FF2 ($X^2 = 3.00$; $p < .10$). The combined judgments favored FF2 ($X^2 = 6.00$; $p < .02$). A second study with different Ss was carried out in the fall of 1965 and involved 4 pairs of actors and 6 pairs of judges. On lines, 11 judgments favored F1F, and 37 favored FF2 ($X^2 = 14.08$; $p < .01$); and on expressions, 14 favored F1F, and 34 favored FF2 ($X^2 = 8.32$; $p < .01$). There were no significant differences in judgments between F1F and FF2 judges. The caution has to be kept in mind that in the role playing this effect was produced by about 7 out of 10 FF2's generally being judged superior to their F1F partners.

In sum, FF2 showed a higher preference for acting as judged by participation and a higher competence at it as judged by role playing—in these experimental settings. In a further study in 1967 involving professional actors in four separate roles, ten members of the four female two-child sibling positions (F1F, F1M, FF2, MF2) were asked to write out an account of what was occurring in the roles. We were interested in which of the groups would most identify with these enactments. The average number of words used by each group to describe each dramatic situation is indicated in Table 8.4. As judged by the number of words used, the FF2 again showed more interest. It is clear, however, that the results are restricted to the FF2 above. They do not hold for the MF2.

We may conclude that there is some support for the thesis that non-firstborns (in this case FF2 only) have greater interest in and competence at acting. This in turn may be taken to support the view that

TABLE 8.4 Average Number of Words Used in Response to Dramatic Situations by Four Two-Child Female Siblings

Roles	F1F N = 10	F1M N = 10	FF2 N = 10	MF2 N = 10
Getting up in morning	50	57	76	48
Restless behavior	69	76	82	56
Preparing for date	50	62	85	54
Being lovelorn	86	91	122	77

non-firstborn find greater satisfaction in opportunities for novel role enactment.

Following this same line of thought, our next study was done with school children, again in search of correlations between role playing interest, competence, and sibling status (Sutton-Smith & Rosenberg, 1967). After exploring various possibilities, children in Grades 2 through 6 in a private day school (upper socioeconomic status) responded to the following sociometric test ($N = 164$).

Instructions for Acting

We all see each other play-acting and making-believe. We all do this differently. Here are a number of ways you can describe other people when they are play-acting. Think which descriptions best fit the children in your class, including yourself. Put a check mark opposite each person's name if the description fits. You may put no checks, one check, or several checks opposite a person's name if all the descriptions fit.

1. This person is always self-confident. He acts without worrying about others looking at him. (CONFIDENT)
2. This person doesn't like to play act. He never does it unless he has to. (WON'T)
3. This person gets embarrassed and self-conscious at having to play act. (EMBARRASS)
4. This person always acts just like himself. He does not change when playing different parts. (ACT SELF)
5. This person is able to act quite different people. He is a real actor. (ACTOR)
6. This person is a funny actor; everyone laughs. (FUNNY)
7. This person makes believe so well that he forgets his ordinary self. (MAKE-BELIEVE)

Prior to the administration of this sociometric device, intensive role-playing experience had been conducted in two of the grades during the previous six months (Grades 2 and 6). While we began simply in the second grade with children guessing the mimes which they enacted for each other, the upper school class began very directly with some of Moreno's techniques such as role substitutions (one person acts the role of another); role reversal (two who have just acted are asked to reverse their roles); mirror technique (children play the role the way they believe that someone else has just played it).

A favorite play with the Second Grade was "adding families." Thus a son would be arguing with his mother over pocket money, and at a particularly dramatic moment we would add a grandfather to the scene and have them all continue from there; then add a sister, and so on. One technique especially indicative of a child's capacity for dramatic play involved having him play two roles at the same time, being alternately each character. At the end of the year the game of "confusion" was developed by the Sixth Grade and is some indication of their skill. It was a contest in which the children played against each other in two's, switching their own roles at random in an attempt to confuse the other player. The winner was chosen by group acclaim.

The importance of emphasizing this special training is that in the results significant findings occurred primarily in these two grades that had had the role experience. The role-taking itself apparently altered the perception of these children so as to permit a more veridical sociometric perception along the above lines to take place. The intercorrelation of these sociometric traits in the two grades is indicated in the Tables 8.5 and 8.6.

In both grades, although the sociometric items confidence, real actor, funny, and make-believe co-vary, these items relate negatively to the other two items: embarrassed and always acts self. There is some intrinsic support here for the notion that the good actor is indeed the one who can take many roles.

In the study of sibling status, the item "real actor," which showed among the highest relationships to the relevant variables, was used as the index of dramatic competence.

Sociometric attributions were converted to standard scores, and the scores of the two classes at each grade level were then combined. Chi

TABLE 8.5 Second Grade Sociometric Correlational Matrix (N = 34)

	1 Confident	2 Em- barrassed	3 Acts Self	4 Real Actor	5 Funny	6 Makes Believe
1. Confident						
2. Embarrassed	−.59**		.60**	−.82**	−.65**	−.82**
3. Acts self	−.21			−.56**	−.38**	−.61**
4. Real actor	.59**				.73**	.71**
5. Funny	.48**					.43*
6. Makes believe	.47**					

All 2-tailed tests, df = 32.
* p < .05.
** p < .01 unless otherwise noted.

TABLE 8.6 Sixth Grade Sociometric Correlational Matrix ($N = 39$)

	1 Confident	2 Em- barrassed	3 Acts Self	4 Real Actor	5 Funny	6 Makes Believe
1. Confident						
2. Embarrassed	−.81**		.27	−.72**	−.72**	−.65**
3. Acts self	−.47**			−.55**	−.53**	−.49**
4. Real actor	.65**				.78**	.64**
5. Funny	.68**					.56**
6. Makes believe	.60**					

All 2-tailed tests, $df = 37$.
** $p < .01$ unless otherwise noted.

square comparisons of the upper half versus the lower half of sociometric distributions (divided at the mean of the transformed score) were run against sex differences, sibling differences, and the effects of previous role practice in Grades 2 and 6. Males were perceived as better actors than females in the role-practice classes (Grade 2: $X^2 = 5.65$, $p < .02$; Grade 6: $X^2 = 1.99$, $p < .20$; Grades 2 plus 6: $X^2 = 6.77$, $p < .01$), but there were no significant differences in the other classes, though the combined scores of all classes also yielded a significant difference in favor of males ($X^2 = 3.83$, $p < .05$). Girls with brothers were perceived as better actors than girls with sisters across all classes ($X^2 = 4.31$, $p < .05$), an effect again produced mainly by the girls with brothers in the two role-practice classes (Grades 2 plus 6: $X^2 = 2.56$, $p < .10$). We should mention that we have replicated this finding that girls with brothers are superior to girls with sisters in a more recent unpublished study ($N = 70$) in which the judgments were made by an expert in theatre.

The present sex differences in favor of males appear to contrast with the fact that "dramatizing" in schools is generally more acceptable to females than males. It contrasts with Bower and London's failure to find sex differences in their Dramatic Acting Test (1965) and it contrasts also with our more recent study in which a theatre expert, echoing common opinion, judged the girls to be superior to boys. It needs to be remembered, however, that in the present study (Sutton-Smith & Rosenberg, 1967) the judgment about superior acting was being made by children who are more appreciative of the exaggerated and ludicrous character of boys' acting than most adults would be. Boys show more novelty in both drama and play. For although boys usually appear to be more squeamish about classroom acting than girls, their free play acting is generally of a more fantastic character (cowboys) and less realistic than the free play acting of girls (house). The encouragement and example of the male investigator was, of course, another possible reason for the judged superiority of the boys in those classes where there was role practice, though it could be argued against this interpretation that the same directional effect was observable in the other grades also. The superiority of girls with brothers over girls with sisters has already been noted in a variety of more conventional tests of ability (Koch, 1954; Schoonover, 1959), and the suggestion has been made that brothers are more challenging and stimulating because they are more prestigious and vigorous siblings. Still, these results are not in accord with the previous expectations that later born would be better at role enactment because of greater flexibility and of greater opportunity to model after a variety of others. Whether this failure is due to the instrument used for measurement or to the upper status sample (males may actually be more efficient in such a sample) is not at present clear.

Clearly the evidence for later-born flexibility and empathy is most tenuous, even though it does seem to fit expectancies drawn from a modeling hypothesis. It could be that because the role modeling for later borns is, in general, less clearly shaped and supported, their behavior as a whole is simply less predictable.

CONCLUSION

Once again most of the later-born behavior has been rationalized in terms of relationships to other children (popularity, empathy), just as it was in the study of sex role and power. Undoubtedly the parents must play some role in all this, but if so, there is little obvious sign of it in these data, though it is true the data are not specifically arranged for that purpose. This situation may be changing (Klaus & Gray, 1968). Contrariwise, the firstborn's role has been explained in this chapter largely in terms of modeling after parents, though a supplementary emphasis has been given to their experience of role surrogation and power over younger siblings. Some of the variables discussed throughout this work to this point are shown in Table 8.7.

The later born give us the paradox of greater autonomy and flexibility, on the one hand, and more consistent copying of siblings, on the other. Still, although the firstborn react to their parents by developing traits (affiliation, achievement) which make a success of the relationship, later born develop much more "noise" in their relationships with older siblings. Their externalized and counteractive behavior is as salient as their imitation. They copy, but they also disaffiliate and protest.

If it is true that the key qualities for the firstborn in their hierarchical interactions are the complexity of the model, the guidance provided in setting up standards, the partial reinforcement, and the long-range achievement that is required to achieve these adult levels, then the later born have by contrast only the standards of a slightly older sibling which

TABLE 8.7 Schematic of Variables

Pattern of Relationship and Influence	Firstborn Reactions
1. *Parental*	
a. Operant behavior (high expectancy and inconsistency)	Affiliation-conformity Achievement Lower self-esteem; anxiety
b. Modeling	Verbal inference Conventionality-conscience
2. *Siblings*	
a. Role surrogation	Nurturance
b. Role power	Dominance; intolerance

Pattern of Relationship and Influence	Later-born Reactions
1. *Parental*	
a. Relative neglect	Autonomy; lower anxiety
2. *Parent-Sibling*	
a. Plurality	Role-taking flexibility Empathy; nonverbal inference
3. *Siblings*	
a. Plurality	Sex role modeling (male and female); counteraction (male) Counteractive aggression

are well within reach. The older child, unlike the parent, may provide very little guidance. While there is evidence that early born have more verbal interaction (and presumably guidance) from parents, no studies focus on the amount of guidance or instruction given by older siblings to younger ones. There is some slight evidence that older siblings mediate parental attitudes rather poorly to the later born (Fauls & Smith, 1956; Strodtbeck & Creelan, 1968). We have offered evidence that without such verbal guidance the later born are forced to pay closer attention perceptually to the doings of their elders. With respect to achievement, it may be supposed that the later born often have the experience of being able to perform as well as the firstborn. Whereas the first children must strive continuously to reach highly valued but relatively unachievable objectives, the second children may gain greater amounts of immediate closure.

If these various contrasts are to be of much research value, then, it is necessary to attempt to reconstruct the type of experimental situation which will parallel the "Pattern of Relationship and Influence" on the left-hand side of Table 8.7. If we can correctly hypothesize the constituents that compose the interactional system within which the subject normally reacts in affiliative or aggressive ways, then those latter responses should be forthcoming.

One such situation might be supposed to represent the role sector: firstborn child and parents (from the hierarchical matrix given above). Let it be called *primary modeling*. The experimenter is authoritative. His requirements are ambiguous and complex. His guidance is explicit and verbal. His reinforcements are alternatively encouraging and critical. It takes considerable persistence and a long time period to accomplish his assigned goals. The prediction is that firstborn will feel more anxiety, seek more reassurance, and persist with greater success than later born. The largely positive results of the Schachter-derived affiliation-conformity studies of Chapters 6, 7, and 8, might be considered to rest on their imperfect reproduction of these conditions. Another situation, to be called *secondary modeling*, would be meant to reproduce some of the conditions of the later-born relations to older siblings. Here the model is a powerful peer; his behavior is explicit and he gives little verbal guidance. Tasks and rewards are immediate. Here the later born are expected to make the stronger response as they do in the Stotland studies mentioned earlier, and as they may well do in many of the Bandura studies of modeling, which often have this character. In passing, we might note that there seems to be some sense in which operant learning is more relevant to the lives of firstborn children, and modeling or observational learning to the lives of the later born. Or, at least, this would follow if the reconstructions of the previous chapters can lay claim to substance. Conceivably, in a choice situation, each sibling set would show preference for learning according to their preferred modality.

Other experimental situations might be contrived to incorporate several important social others of varying power, thereby attempting to reproduce upward and downward the hierarchical relationships of the firstborn. The subject might be required to model after and react to power superiors with the characteristics outlined in primary modeling above, and also to react to power inferiors with the characteristics in secondary modeling given above. Several studies that incorporate some of these alternatives have already been quoted (Gordon, 1966; Arrowood & Amoroso, 1965). They are worth reproducing in more detail here. We shall refer to them as *hierarchical modeling* experiments.

Arrowood and Amoroso, using 92 female undergraduates at Toronto, had the subjects read a case study and make ratings upon it. They were then led in one condition to believe that their ratings conformed to the group consensus, and in another condition to believe that their ratings deviated from the group consensus. In addition, group members were given communications ostensibly from other group members by which they also became aware of whether these other members were conformers or deviates. Finally, they were given the opportunity to rank all members sociometrically in order of their preference for their being in a future discussion group. The findings were that the firstborn felt less confident about themselves, and more often rejected themselves from the future group if they were deviates. Firstborn deviates changed their opinion toward the modal group opinion more readily than did later born deviates—all of which supports earlier research. But the really important finding was that firstborn who thought they were conformers rejected deviates more readily than did later-born conformers.

Gordon (1966) also found that when firstborn felt certain of their opinion, they were more interested in convincing others that their own opinions were correct.

A very strong example of the way in which internal dispositions interact with situations to cause differential success or failure has been provided by John E. Exner (1969) in a series of studies of Peace Corps Volunteers. While a number of variables contribute to their success or failure, the direction of the ordinal position contribution depends on the interactional context. In relatively open ended situations requiring innovation and an externalized existence (much of Micronesia), later born are more successful. But in a relatively formal situation requiring conformity to a hierarchically organized structure (much of East Asia), only and firstborn volunteers are more successful.

It has been argued that each set of siblings (first and later) is caught in a particular interactional web, and that the various influences act to produce characteristic patterns of response. The relevance of the schematic of variables (Table 8.7) is that it implies that in predicting differences between siblings of different positions the relevant social matrix should be kept in mind. Although there was some evidence from the study of the relationships between achievement and conformity that achievement becomes increasingly autonomous as a response system, such independence should not be taken for granted. It would normally be expected that insofar as any given set of social conditions approximate those of the matrix, the greater would be the probability of transference of the learned reactions. Thus we suspect that firstborn are more likely to be affiliative when relating to authoritative persons, who, like parents, are more complex than the subjects themselves who rely on verbal and conceptual guidance and who expect a great deal of the subject, but are somewhat inconsistent in their encouragement and criticism.

One might argue in these terms that firstborn might be drawn into education by the way in which, first their teachers, and subsequently their professors (often themselves first born), parallel in their own behavior the original family matrix conditions. Perhaps, facetiously, the situation is even modeled by the conceptions of the worshiper and

God in industrial and Protestant society. Both God and his standards are only partly explicit. His help can be solicited for cognitive and affective guidance; in hymns one sings of one's own uncertainty and low esteem but also of ultimate affiliation with God; in prayers one promises conformity to God's standards; but in works one achieves beatification. This is to say that in his uncertain universe, the Great Partial Reinforcer maximizes affiliation, conformity, and achievement.

In sum, there is already much evidence that experiments which are representative of sectors of the interactions in which siblings of different status find themselves can reproduce in those siblings the expected responses. There are yet other sectors not dealt with here (for example, the later born's relationships to a plurality of power types), but enough evidence has been presented to imply that there is some value to thinking about sibling status conjointly in terms of the *structures of interaction* mentioned above, as well as in terms of mediating types of learning (operant, modeling, and counteractive) and the cultural contents (affiliation, aggression), when one is attempting to make predictions about future behavior.

What is lacking in this quite schematic account is a detailed analysis of what has been meant by counteraction as a type of learning. While there are concepts in the literature of analogous character—"reaction formation," "masculine protest," "compensation," "negative identification," "dialectical response," "adaptation level,"—the present data allow us only to point a finger to the reality. Undoubtedly the category of "counteraction" conceals more than it reveals. Perhaps, for example, the father with the two daughters or the brother with the two sisters do not really "counteract" their feminine influence. It is possible that their behaviors are actually evoked by the females themselves, who are more responsive to such behavior. Where "maleness" is in short supply, it might be reacted to more strongly (either positively or negatively) by the female family members. Thus intentionally or otherwise the females may reinforce that behavior in the male repertoire. Similarly the later-born child may find he can get, what is to him, a more satisfying "reaction" from his older sibling by harassment rather than by physical attack. With the former he has some success, with the latter none. Again, it is the firstborn reaction which shapes up his "counteractive" response. In both cases, that is, we might subsume this "counteractive" category to a variety of operant conditioning, always admitting, of course, that this accounts only for the shaping up of these responses, not for their origin. Still, lacking any specific evidence, we prefer at this time, to point a finger at this category of data that we call "counteractive" rather than to offer a solution.

A further deficiency, and a more important one, in the present discussion and throughout the previous four chapters has been the almost complete lack of attention to specific ordinal positions and to developmental phenomena. This has been caused by a paucity of these considerations in the research reviewed. These matters are taken up again in the next chapter.

CHAPTER 9

consistencies and transformations in sibling status

The preceding four chapters have with some exceptions considered the literature on firstborn versus second born, regardless of the type of precautions mentioned in Chapter 2 and spelled out in some detail in Chapters 3 and 4. Unfortunately in the achievement, affiliation, and conformity literature there are few references to specific ordinal positions, to the sex status of the siblings, or to the age spacing between them. Usually there is some control for family size and socioeconomic status. From the little information available it seems that most of the contrasts made probably hold best when the age spacing is between two and four years, when the firstborn and later born are of the same sex, and when the subjects are of middle socioeconomic status. Actually there is very little information on the effects of varying socioeconomic position, though it seems likely that differences in such status as well as subcultural differences of other sorts might considerably modify the generalities presented in these chapters (Rhine, 1969). The data of Rosen (1964) on the agreement of children and mothers over achievement values, cited in Chapter 5, show such differences. There are examples of subcultural differences in the literature. Thus Becker and

Carroll (1962) asked preadolescent boys in a Chicago playground to make judgments about the relative lengths of various lines, after hearing judgments made by other boys. These latter boys were the experimenters' accomplices and made false estimates by prearrangement. The question raised by Becker and Carroll was whether firstborn conform more in this situation than do later born. In the case of this sample of American boys, it was found that the firstborn yielded more to the falsifying influence of the accomplices than did the later born. But the Puerto Rican boys in the sample did not show this ordinal position difference. They were all highly conforming.

Similarly, a series of studies involving assessments of both affiliation and reactions to frustration has yielded some interesting differences between students at Yale and students in New York City. Sarnoff and Zimbardo (1961) reported that under threat of shock, Yale firstborn showed a greater preference for spending their time with others while waiting for the experiment in which the shock would occur than did Yale later born. Gerard and Rabbie (1961), however, found that their Brooklyn College students were not similarly affected; indeed there was a tendency for the second born to be the ones who showed such affiliation tendencies. Analogously, Glass, Horowitz, Firestone, and Grinker (1963) pointed out that under circumstances of experimentally induced frustration at Yale, firstborn displayed more annoyance (Dittes, 1961), but at New York University second born showed more annoyance. While there are a number of differences in methodology between these various studies which present the possibility that the contrasts between Yale and New York City students are more apparent than real, if taken at their face value, they seem to imply cultural differences between the Yale and New York students. Glass et al. state: "We suggest the consideration of sociocultural factors in any effort to interpret these differences, since we do not regard birth order as a psychological variable" (1963, p. 194). As the Yale firstborn students appear to be more sensitive both to threat and to annoyance, we are led to wonder whether the relative selectivity of Yale and the higher socioeconomic standing of its students are in some way connected with these birth order differences, whereas the relative neglect of second borns in the predominantly Jewish New York City culture may be leading to a similar response for functionally dissimilar reasons.

Interpretation is perhaps facilitated by a recent study in Israel (Amir, Sharan, & Kovarsky, 1968), in which firstborn of Western nuclear families (like the Yale students) were compared with firstborn from Middle Eastern extended and nuclear families (like New York students), with the finding that in the Western group the firstborn were significantly underrepresented among those in officer training, and in the Middle Eastern group they were significantly overrepresented among those in officer training. The difference is explained by these authors as that between Eastern firstborn who have the traditional expectation of preferential status and those who do not. Where the firstborn has the expectation of preferential status in his family, then the anxieties and apprehension with respect to physical danger associated with firstborn status throughout this book do not exist.

> From biblical times Jewish tradition has accorded the first born male a position of primacy. His preferred position was given palpable recognition by the inheritance law granting him double the portion received by his younger siblings. . . . One would anticipate that the first born male's acceptance in the Middle Eastern family would counteract or neutralize the anxiety-arousing effects of the typical mother-firstborn relationship. . . . " [Amir et al., 1968, p. 272].

While it is not clear from this report whether these preferred firstborn are initially anxious, or if such anxieties exist, how they are modified or directed, we are presented with the distinct possibility that the firstborn anxiety-affiliation-achievement syndrome presented throughout this work is a vestigial state; that is, a complex of attributes left over from this earlier order of primogeniture, and further evidence of the decay of siblingship as a status variable.

Still these are but speculations, and do little but impress us with the importance of further examining the family or ethnic conditions which differentially affect the generalizations that we have derived from the previous chapters.

The major concern of this chapter is with a variety of *developmental variables* which might be considered to modify the picture already presented. In the search for consistent differences between firstborn and later born these have been neglected except for age changes in the account of relationships between affiliation and achievement. The procedure will be to discuss some general developmental factors which might be considered as playing a part in contributing to changes in sibling differences over time, and then to take up each of the sibling statuses separately and trace what is known of its characteristics as they change with age.

THE DURATION OF THE RELATIONSHIPS

As we do not yet have sufficient studies of the effects of siblings and parents upon each other when the relationships endure, we are not in a strong position to talk about their effects when such relationships do not endure, or occur only for certain age periods of development, as might happen as a result of mortality, separation of family members, the introduction of new foster children, and so on. We do, however, have one study in which such a mutuality of effects was considered. Here we were concerned with the effects of siblings upon each other when the fathers were absent during different periods in the children's lives (Sutton-Smith & Rosenberg, 1968). The study focused upon the relative effects of father presence or absence, and of sibling presence or absence, upon the cognitive abilities of family members of different sibling positions in single-, two-, and three-child families. It is conceivable, for example, that the effect of having both a father and a brother is quite different from the effect of having a father without a brother or a brother without a father. Although there have been studies of the deleterious effects of father absence upon children (Nash, 1965), there has not been any consideration of these father absence effects in families of varying sibling composition. The question raised was whether father absence is equally deleterious to children from different sibling statuses and family sizes. In raising this question, it made sense to argue that the effects of father absence would depend both on the length of his absence and upon the age period in the child's life during which he was absent. The antecedent variables in this study therefore were, variously, the length of father's absence, the child's age at the time of absence, and the type of sibling composition of which the subject was a member. Subjects in the 8 two-child sibling positions were included. Also included, at least as a group, were the members of the 24 categories of the three-child family. These various sibling groupings made it possible to compare subjects of first-born and second-born status, subjects with like and opposite sex siblings, and subjects from only-child, two-child, and three-child families.

The consequent variables chosen for analysis were the subjects' scores on the American College Entrance Examination (ACE). This is a test of quantitative and language abilities. An earlier study by Carlsmith (1964) had already shown that father absence has an effect upon mathematical and verbal aptitude scores taken at the college level.

The scores on this test for the various groups on quantitative (Q), language (L), and total (T) scores are indicated in Table 9.1. Since Ns were too small for MM2 and FM2, these are presented in parenthesis. For the entire sample it is apparent that father absence effects were dramatic. For the majority of the sample, Q, L, and T scores on the ACE were lower for father-absent subjects than for father-present subjects. This effect obtained more uniformly for males than females. In addition, these effects increased with family size. Not shown in this table was the rather surprising finding that father absence effects were stronger when the father was absent between the children's ages of five and ten years than between zero and five or ten and fifteen years. The total length of the father's absence (over two years) made no difference. In a related study of fathers on night shift work it was found that their relative absence had an effect on the children under 10 years, but not on those over 10 years (Landy, Rosenberg, & Sutton-Smith, 1969). More important,

TABLE 9.1 Comparison of Median Scores on ACE for FA–FP for Total Sample

	Father Absent				Father Present				Differences (t tests; p, level)		
	Q	L	T	Ns	Q	L	T	Ns	Q	L	T
M	53	54	47	(21)	56	45	48	(35)	—	.20	—
F	53	66	61	(46)	71	71	67	(51)	.05	.20	.10
M1M	49	38	44	(20)	67	54	56	(38)	.10	.10	.20
M1F	49	33	35	(17)	71	58	67	(34)	.01	.001	.001
MM2	(68)	(53)	(60)	(6)	62	43	54	(33)	—	—	—
FM2	(44)	(42)	(36)	(6)	72	63	68	(44)	—	—	—
F1F	56	67	61	(40)	60	61	56	(43)	—	—	—
F1M	49	69	60	(36)	71	61	68	(68)	.02	—	—
FF2	64	58	68	(15)	59	63	59	(48)	.20	—	—
MF2	60	70	67	(17)	71	61	66	(59)	.20	—	—
2 child males	49	38	41	(49)	71	56	63	(149)	.01	.001	.001
2 child females	56	65	61	(108)	67	63	63	(217)	.10	—	—
3 child males	57	30	39	(27)	69	48	58	(136)	.001	.001	.001
3 child females	53	65	61	(44)	66	63	66	(172)	.001	.001	.001
Total Ms	53	49	44	(97)	71	55	63	(320)	.001	.001	.001
Total Fs	56	65	61	(198)	67	54	66	(440)	.001	—	.20

SOURCE: B. Sutton-Smith, B. G. Rosenberg, & F. Landy. Father-absence effects in families of different sibling compositions, *Child Development*, 1968, **39**, p. 1216.

NOTE: Q = Quantitative Scores; L = Language Scores; T = Total Scores.

from the present viewpoint, the evidence in the table also shows fairly clearly that in the two-child family the greatest differences between father absence and father presence are produced when the child has an opposite-sex sibling. Boys with brothers and girls with sisters are least affected by the fathers' absence. There are some sibling circumstances, therefore, in which the presence of a sibling can mask the effects of a parent's absence, and there are some *periods in development for which this masking effect is more critical than others.*

AGE OF SUBJECTS

Koch (1955a) selected her particular age group of six-year-olds because she felt they would still show the dominant effect of family influences and yet be old enough to communicate with the investigator. It is most evident in her results that she was dealing not just with the eight types of sibling positions, but with siblings whose relationships were those between six-year-olds and two-year-olds, six-and-four, six-and-eight, six-and-ten, and so forth. And these particular age relationships appeared to be just as important in many of her outcomes as the character of the siblings. Thus, Koch says younger children who are under two years and of the same sex were threatening to the older sibling. If they were younger and of the opposite sex, however, the older sibling was less perturbed by them. On the other hand, the older sibling was more disturbed by the opposite-sex younger sibling if he or she was close in age. It would appear that the younger sibling of the same sex doesn't challenge one's sex role identity (indeed reaffirms it), but in getting more mother attention perhaps invokes anxieties of a more oedipal character. Contrarily, the opposite-sex very young sibling can never have the same relationship to the mother as the subject, but if the sibling is older and of the opposite sex it can cast doubt upon the validity of the sex role currently being followed by the subject. In saying these things we are, of course, mixing interpretation with data. The important point is that Koch's results are interpenetrated by the age relevance of the subject-sibling comparisons. Although both her firstborn and second born were six-year-olds, in the case of the firstborn whose siblings were under six years of age, she was really dealing with the effects of the nuclear family and primary ties. But in the case of the second born, she was dealing with six-year-olds whose older siblings had moved out to the world of secondary ties, peers, and teachers. So we have a contrast, not just between ordinal positions, but between six-year-olds whose siblings tie them back into the primary family and six-year-olds whose siblings lead them out into the secondary groups. The differences between them thus have much to do with whether the ages of the siblings mediate primary-group or secondary-group influences.

Koch's major finding that the first children were more concerned with parents and the second children with siblings may well have had much to do with such a pattern of mediation. This does not deny the relevance of the sibling differences she revealed, but it does show the way they were complexly interwoven with the age of her subjects.

Again, in our own college data, we have found that the scores on the Gough Femininity Scale (1952) derived from the children and parents of a specific sibling grouping (say, F1F–FF2) depend on whether, when we are getting those data, one of the sisters is still at home or one has now left home and is married. When there is such an age shift in the family relationships, it is reflected in how the members respond to an inventory. Fathers, for example, respond in a more feminine way when there is only one girl at home as compared with two. When they have two daughters at home, the fathers respond in a more masculine way (Chapter 3). More substantial evidence bearing on the way in which the age of the subjects enters complexly into sibling differences is given by the only longitudinal study undertaken on sibling effects, namely the study by Lasko, who in 1954 reported a study of differential parental behavior to first and second children. One of her findings was that the mother tended to treat the children more rationally and democratically if they were closer together in age than if they were widely spaced. As Lasko notes:

> One can only theorize as to the selective factors which might be operant: it may be that mothers who subscribe to modern tenets of child care also tend to have children closer together than do the so-called old-fashioned mothers. Or, it may be that two children close together in age are more satisfactorily handled by these methods because of the similarity of their general developmental stage needs [1954, p. 113].

In addition, Lasko found that with the advent of the new baby, the mother found the four-year-old easier to handle than the three-year-old, and was harsher toward the latter. Here the age stage characteristics of the four-year-old made it easier for the mother to have another baby. The obstreperous characteristics of the less mature three-year-old perhaps did not. Again at five years of age, on the verge of school entrance, acceleratory pressure was applied to both firstborn and second born.

In a cross-sectional study of our own (Sutton-Smith & Rosenberg, 1965a) comparing the anxiety scores of subjects in two-child families and using Koch (1956c) data for the six-year-olds and our own data for the preadolescent and college level subjects, we found that anxiety was higher among the firstborn at age six, but was highest among those with older sisters at ages eleven and nineteen. It appears in these data that the critical birth order factors interacting with anxiety change with age. Perhaps the firstborn is made anxious at age six by the barrier between him and his mother presented by the second born, but the subjects with sisters are made more anxious at age eleven and nineteen because anxiety is more typical of the female than the male sex role, and they are influenced by their sister.

These are but a few illustrations of the way in which age difference expectations and competences on the part of the children interact with influences peculiar to sibling status.

AGE OF MOTHER

It is fairly well established that younger and older mothers show higher incidences of stillbirths, mortality, mental deficiency, and mal-development in their offspring (Broverman & Klaiber, 1968). It is not impossible, therefore, that more subtle effects of mother's age at the time of pregnancy may be confounded with birth order. After all, first-born will generally have younger mothers than later born. Broverman and Klaiber have demonstrated that older and younger mothers give rise to offspring who are, as they term them, weak automatizers. Autom-atization is measured by the rapidity with which a subject can reduce a routine task to a habit level. Characteristically, the subject has to name a series of pictured objects arranged in a random sequence. The high automatizers carry out this simple perceptual-labeling task with high speed. The low automatizers continue to fumble along haltingly. In other studies low automatizers have been shown to react less adequately to stress, to be more distractible, to attain lower-level occupations, and possibly to be less androgenized (less masculine) at the hormone level. While no significant relationships have been obtained between such mother's age phenomena and birth order, the results do indicate the subtle way in which hormonal differences associated with mother's age might have the possibility of affecting the offsprings' psychological processes.

On a more obvious level, younger mothers have more stamina and vigor than older mothers. One speculation in the literature is that they are also more anxious and uncertain about their child-training pro-cedures, and that this has the effect of inducing anxiety in their off-spring (Chapter 6). Rosen (1964) interviewed both mothers and sons of ages 8 to 14 and compared what mothers of different ages did with their children, and the extent to which mother and son values were similar. His younger mothers were between 20 and 39 years; his older mothers were over 39 years. He found that younger mothers said they trained their sons earlier, but older mothers tended to make more use of psychological disciplinary techniques. There was an interesting inter-action with social class. Mothers in the middle class, whether young or old, were more similar in value to their sons, but mothers in the lower class were similar to their sons in value only if the mothers were older. Shrader and Leventhal (1968) found that younger mothers reported more problem behavior in firstborn than in later born. Older mothers did not make this distinction.

In her book on twins, Koch (1966) found an even more subtle interaction between age and other variables. In matching her various groups of twins she found that the mothers of her dyzygotic (nonidentical) twins tended to be older than the mothers of her monozygotic (identical) twins and apparently had had more difficulty in conceiving. While there were no differences in age at marriage, the mothers of the dyzygotics had taken a longer period before having their first offspring (1966, p. 172).

FADS IN CHILD REARING

Any developmental study must take into account not only the developmental status of its subjects (in this case conveniently epitomized by age levels of siblings and mothers) but also the historical status of the behaviors being considered. In the past 50 years, for example, there have been many changes in child-rearing practices, particularly in the direction of more permissiveness. It is not inconceivable, therefore, that if a mother through the course of her own development (ages 20 to 40) undergoes any educational upgrading (comes more in contact with the advice of the experts), this might affect her practices across the birth orders. A change of this sort might make her less inclined to use physical disciplinary techniques and more inclined to use psychological techniques as she gets older. Such a difference is found in Rosen's data (1964) between younger and older mothers and in the data on physical punishment (Clausen, 1966). Unfortunately, these are not studies of the same mothers over time. Studies of the same mothers, such as that of Lasko (1954), indicate, if anything, a contrary trend toward their becoming more restrictive and arbitrary with the later birth orders.

Once again, however, we are concerned here only with establishing the relevance of such variables, which is, at this time, about all that the evidence will show. What we have is prima facie evidence for the view that siblings as they age are likely to have transformations introduced into their development as the result of these or other factors. Our first evidence that this is actually the case will derive from a reanalysis of several of the affiliation-conformity studies in which analysis was made separately for the eight two-child sibling statuses.

DEPENDENCY IN TWO-CHILD FAMILIES

There are four studies in which the eight two-child statuses were dealt with separately and in which the variables were analogous to dependency. This review is dealt with separately before the assessment of each of the specific sibling positions because it bears importantly on the largest group of current experimental ordinal position studies—those derived from Schachter's original findings—and also permits some conclusions with respect to his developmental thesis.

The studies to be reviewed first are those by Koch (1955a), by Bragg and Allen (1967), and by Sampson and Hancock (1967). Koch had her teachers rate her subjects on a number of their attitudes toward adults. Four of the traits that she used may be taken to be similar to the "affiliation" and "dependency" that was discussed in Chapter 6. They are "friendliness to adults," "affectionateness," "tendency to seek for adult attention," "tendency to appeal to adults for help." As in Haeberle's results (1958), the girls scored higher on these traits than did the boys. Affiliation was a more sex-typical response for girls, and girls were perceived by Koch's teachers as showing more friendliness, and help-seeking.

In Table 9.2, the 8 two-child positions in Koch's study have been ranked in the order of their ratings on these four traits when pooled together. In addition, the rankings of two other studies in which contrasts between different types of ordinal position on variables which are theoretically consonant with those discussed in Chapter 6 are also included. Certain characteristics are noteworthy. In particular, there is considerable consistency across these rankings with respect to the subjects occupying the extreme positions. In Koch's study, the FF2 (second-born girl with an older sister) and F1M (firstborn girl with a younger brother) received the highest scores, and MM2 (second-born boy with an older brother) and M1F (firstborn boy with a younger sister) received the lowest scores. The Bragg and Allen study (1967) is the most important because it was an experimental conformity study involving group pressure from accomplices.

Again, FF2 and F1M were high conformers, and MM2 was a low conformer. Sampson and Hancock's self-report data from the Edwards Personal Preference Schedule again gives the FF2 and F1M the highest scores, and the M1F and MM2 the lowest scores (1967). These consistencies in responses across the three studies involving different though theoretically related variables at different age levels are impressive. But they do raise important and at present unanswerable questions about the successful results quoted in the introductory review of experimental studies (Chapter 6). The Bragg and Allen study using the experimental conformity paradigm ranked both types of firstborn males (M1M and M1F) higher than both types of second-born males (MM2 and FM2), but not firstborn females over second-born females. This is consistent with the results in the review presented earlier, which held for males (9/2) but not for females (2/5). Unfortunately, there have been no

TABLE 9.2 Specific Ordinal Studies of "Dependency"

Investigator	Description of Variable	Rank Order							
		1	2	3	4	5	6	7	8
Koch 1955 (6-year-olds)	Attitudes to Adult (Ratings)	FF2	F1M	MF2	FM2	M1M	F1F	MM2	M1F
Bragg & Allen 1967 (College)	Conformity (Experiment)	FF2	F1M	M1F	M1M	MF2	FM2	F1F	MM2
Sampson & Hancock 1967 (Adolescent)	N Affiliation (EPPS)	F1M	FF2	F1F	MF2	M1M	FM2	M1F	MM2

anxiety-affiliation experimental studies with a similar breakdown by ordinal position, although the Sampson and Hancock data, if representative, could give successful firstborn versus second-born differences because firstborn have rank positions 1, 3, 5, 7, while second born have rank positions 2, 4, 6, 8. Any slight skewing of the sample in favor of particular types of firstborn or second born could lead to a positive result. The major conclusion from these comparisons, however, is that little can be said about the 34 other studies until they are methodologically revised in terms of these specific ordinal positions.

In the Berkeley Guidance data (MacFarlane, 1938) there is a dimension of *emotional dependence—independence* on which there are ratings for the years 6 to 16. In this study also the F1M and the FF2 ($N = 14$) were rated the most emotionally dependent throughout, and the MF2 and F1F ($N = 15$) were rated as more independent. The greatest contrast was between the relatively dependent F1M and the relatively independent F1F, with the second-born subjects less clearly distinguished. The ratings for the males were more complex, with none of these clear-cut and consistent differences across all age levels, as in the case of the females.

The following section presents various possible explanations for the differences between siblings mentioned above and at the same time suggests new bases for the subsequent study of the interaction between birth order and the affiliation-conformity variables. In addition, it brings together in summary fashion the data on each of the sibling positions in the two-child family, only children, middle born, and last born.

SPECIFIC SIBLING STATUS

Second-born Males of Same Sex (MM2)

Second-born brothers with older brothers (MM2) and second-born sisters with older sisters (FF2) present the greatest consistency of response in the four studies discussed above. The responses of the MM2, being low on all these dependency-related measures, conform to both the Schachter expectations and the preschool findings for second born in general. Being second born, they are expected to be low on measures of dependency, affiliation, and conformity, and that is indeed how they appear in Table 9.2. The responses of the FF2, on the other hand, although consistent in themselves, completely reverse the Schachter-based expectations. Instead of being low on dependency, affiliation, and conformity, as are the MM2, these girls obtain the highest scores on these measures (see Table 9.2).

In order to explain these differences, Bragg and Allen, authors of the conformity study shown in the table, advance a role-taking theory. They suggest that conformity is an appropriate role characteristic for females, but an inappropriate role characteristic for males. In addition, they advance a role-modeling explanation: "A role modelling theory of conformity suggests that a younger child with an older sibling will learn sex role behavior from the older model. An elder male will serve as a model of nonconformity while an elder female will serve as a model for conformity" (1967, p. 2). In these terms, FF2 will be conforming, and MM2 will be nonconforming, which is congruent with Bragg and Allen's findings. The difficulty with this explanation is that it stipulates an additional causal agency for conforming behavior. In addition to parent-induced conformity during infancy there is now sibling-induced conformity. If the assumption is made that both types of causal influence do in fact operate, then it would follow that they work toward a consistent end in the case of the MM2, and toward a contradictory end in the case of the FF2. The MM2, already low on dependency in relation to parents, effectively models after his "nonconforming" elder brother, who is himself relatively unfriendly to adults, according to Koch (1955a). This MM2 gives perhaps the best illustration of modeling after a higher-powered sibling.

In addition, this male presents us with one of the most consistent pictures of development. He is the most masculine, least feminine, and least anxious—on Koch measures at age 6, and our own measures at ages 11 and 19—or he is among those who are. He is least conforming and affiliative on the measures quoted above. He is the most athletic in preferences on the recreational inventory (Chapter 3) and more likely to enter dangerous sports than the firstborn (Nisbett, 1968). He makes higher scores on masculinity and achievement measures during adolescence than do the FM2 (Teepen, 1963). On the Strong Occupational Inventory these boys, along with their brothers, as compared with boys who had sisters, showed the greater preference for the activities of producing, buying, and selling, all of which are a part of the standard economic transactions in this culture. These occupations were life insurance salesman, buyer, real estate salesman, banker, purchasing agent, production manager, farmer, accountant, sales manager, and president of a manufacturing company. Similarly Leventhal (1966) found these MM2 to be more interested in entrepreneurial and management activities than were the MF2. But it was also shown (Chapter 4) that these boys are the most overpowered and put upon of all the younger siblings. Koch's resolution of these differences is as follows: ". . . although bossed, out-stripped and dominated by the mother resenting older one [he] is nevertheless impressed by the skill and achievements of the slightly older siblings" (1955a, p. 32). Overpowered by the older sibling, he models after his masculine traits—an appropriate exemplification, apparently of the importance of the power of the model in inducing imitative behavior. The small Berkeley sample of younger brothers ($N = 8$) yielded some consistencies with this general pattern of data and some interesting differences. As expected, these boys were the least anxious of the total group at ages 6, 10, and 14 years.

But the most interesting data had to do with the reciprocal changes between these boys and their elder brothers (M1M) on the ratings of independence and dominance. Whereas the MM2 boys were above average in ratings of emotional independence from 5 years of age through 9 years; from 9 to 13 years of age they were below average in independence. The scores on independence for the older brothers' group, on the other hand, showed a complementary reversal, being below average until 9 years of age, and above average from 9 to 13 years of age. The MM2 also showed a shift from an average position on the dominance-subordinance ratings from age 5 to 9 years to a very subordinate position from ages 9 to 14 years. It appears from the Berkeley ratings that what has to be explained is the marked drop in independence and dominance of these younger brothers between the ages of 9 and 14 years. That is, while their ratings on anxiety, masculinity, and so forth, remain consistent over these years, they nevertheless show a surge in emotional dependence and submissiveness during the preadolescent period.

The only explanation we have to offer for this partial transformation in these otherwise most consistent of boys is that having modeled after the older boys during childhood (and thus shown appropriate male nonconformity, nonanxiety, independence, and bossiness), when the older male leaves them and goes off to high school (which he will do when they are about 9 years old) and they must make their own way with a peer group in preadolescence, they are much less well equipped than they appeared to be when they had the guidance of their older brother. They resort to the dependence and submissiveness which were also a part of their power tactics repertoire as second-born brothers. During childhood this submissiveness and dependence had perhaps been masked by a display of brother-imitating masculinity. But without the brother for immediate feedback in the preadolescent situation, they fall back on these lesser responses in their repertoire. At age 14 years in the Berkeley data, they once again demonstrate a surge in dominance and independence, returning by age 16 to the high level demonstrated during the years 6 to 9.

Given the small size of the Berkeley sample, there can be no great confidence about these descriptions. On the other hand, it does make some sense that in the traits that can be modeled directly (masculinity, athletics, sports, entrepreneurialism, nonconformity, nonanxiety) there is consistency over the growth period; but in the power characteristics where the subject is both modeling from the older sibling (domineering, independent) and reacting to him (submissiveness and dependence), there should be a mixed expression of these characteristics as the life circumstances shift.

Second-born Females of Same Sex (FF2)

The FF2 is somewhat more complex. Her scores on the dependency measures are not what would be expected of a second born. If she was relatively nondependent on parents in preschool years as she was shown to be, why should she now appear to be more dependent in these child-hood and adult measures? Perhaps the paradox here is not as great as it seems. Being relatively nondependent on parents as a preschooler, she is freer to model after siblings and peers. Her sibling and peer models in early childhood would be older girls. It is modal at that age level for young girls to be dependent on adults. This second-born girl, then, is free to adopt this appropriate sex role attitude, and in Koch's ratings she appears highly adult dependent. During adolescent years, however, as the modal sex role for girls involves less dependence upon adults, she could be expected to be less compliant toward adults. Bragg and Allen's conformity study involving peer conformity is in accord with this finding.

Still, to adopt such arguments is also to imply that the measures of dependency described in Table 9.2 are actually assessing different types of conformity response in the different ordinal subjects, and that the apparent similarities between them may have quite different functional bases.

In other data, as expected, she is one of the most feminine and least masculine of the two-child family female sibling statuses. Koch reports she feels closer to the mother, while her older sister feels closer to the father. In the Berkeley Guidance data, she is less anxious than the older sister during the adolescent years, and generally less independent, less competitive, and less domineering throughout.

Firstborn Males of Same Sex (M1M)

The older members of these same dyads (M1M and F1F) seem to require explanations involving their relationships with parents rather than explanations that rely heavily on the concepts of modeling after higher-powered siblings. By age 6 they will have had several years in which primacy of access to the mother was absent. Following Koch it might be argued that as the younger sibling is of the same sex, then there is probably no great conflict about their own sex role adequacy. There may be, however, increasing rivalry for the solitary opposite-sex parent in these same sibling sex families.

Like the younger brother, M1M shows a highly masculine pattern of interest and occupational preferences over the years on which there are measures. Unlike the younger brother, however, he also has above-average rankings on anxiety at the 6- and 10-year age levels (Sutton-Smith & Rosenberg, 1965a); and in the measures of affiliation and conformity mentioned above, he is one of the most conforming males. The argument was presented earlier that younger brothers, while they do not challenge the older brother's sense of sex role adequacy, create rivalry for the mother (one woman among three males). The M1M's maleness is not in doubt, but his access to the mother is. Thus challenged, he tries harder to please her (affiliate, conform) and is anxious about his relationship with her. Koch's data on six-year-olds seem consistent with such a view. The teachers rated these boys as the most quarrelsome, the most teasing, insistent on their rights, and slow to recover after upsets. These subjects contended in interviews that their mother favored the younger sibling, and they gave unpleasant characterizations of the mother in their Children's Apperception Test scores.

The Berkeley Guidance data appear to confirm this pattern. While the M1M remain above the average on dominance throughout (as would be expected from the chapter on power tactics), from ages 6 to 10 they are well below average on emotional independence, and are high on anxiety and quarrelsomeness. After the age of 10, however, there is a striking reversal of scores, with the M1M becoming above average in emotional independence and competitiveness, being low in quarrelsomeness, and reducing to average levels on anxiety. These ratings strongly suggest that with the arrival of preadolescence there is a considerable decrease in tension for these older siblings. Possibly, their new independence among peers, and their maturing skills enable them to take up a special position vis-à-vis their mother that no longer requires that they compete with the younger brother on his immature terms. When the dyad of MM2 and M1M are considered together, while they both show a consistent masculinity throughout, the older sibling seems to have greater adjustment difficulty before the age of about 10, and the later born after that age.

Firstborn Females of Same Sex (F1F)

According to Koch (1955a) the F1F see themselves in the caretaker role. They like their teachers and are rated as sensitive, feminine, and having good relationships with the younger sibling. Their playmates are mainly female; their models female. They have opportunity to play parent surrogate with a relatively compliant female younger sibling. Nevertheless, they see the mother as favoring the younger one, and themselves as championed more often by the father. The cross-sex data of Chapter 3 seemed to imply that this group was a particularly competent group; and that it might be easier to learn the female role in this mother surrogate position. In the Berkeley Guidance data, they are the most independent of the girls' groups at all ages. In examining this group (M1M and F1F) as well as the previous groups (MM2 and FF2), it is clear that a Schachter-type "dependency" study carried out with M1M and MM2 should succeed, but with F1F and FF2 it should fail.

Second-born Males of Opposite Sex (FM2)

Second-born subjects with older opposite-sex sibling (MF2 and FM2) should acquire many of the opposite-sex traits of the older sibling through modeling after his or her greater power. This means, however, that their own sex role traits will be modified rather than exaggerated as they were in the previous cases of the MM2 and FF2. Table 9.2 shows indeed that FM2 has the most feminine rating among the four boy groups. Insofar as the boy with an older sister becomes more feminine and therefore more dependent, and thus veers away from the expected lower dependency of second born, then the Schachter-derived infant predictions will fail to hold up. But the girl with an older brother, in becoming more masculine or less dependent, should help to maintain expectations that second born will have lower scores on conformity, affiliation, and so forth.

Not surprisingly, in Table 9.2 these two ordinal positions hold scores of an intermediate character. Self-report measures from subsequent years bear out this tendency for MF2 to veer toward masculinity in her responses, and for FM2 to veer toward a more feminine responsiveness than is typical for his sex (Sutton-Smith & Rosenberg, 1965a). Modeling after an older sibling of the opposite sex appears to be a powerful influence against modal sex role expectations. In Koch's six-year-old data (1955b), FM2 is characterized as withdrawn and depressive rather than outgoing and enthusiastic. His older sister, by contrast, is outgoing and enthusiastic. It is a paradox that at age 6 he lags in the very characteristics that he appears to instigate in his older sister. He is rated as low on gregariousness and friendliness, and yet is seen as quarrelsome, exhibitionistic, selfish, and uncooperative with his peers, as well as given to teasing. He is rated as sissy, not tenacious, not ambitious, not competitive, and not insistent upon his rights. Perhaps because of the sister's oppressive influence, these boys reported fewer than usual associations with the older siblings; more than others they did not want to change places with her. If there was to be a new baby in the family, they hoped it would be a male. They reported being favored by the father. In Chapter 4, however, it was shown that they were the most powerful of the second-born siblings as judged by the power level of the tactics used upon the older sibling, and in the Berkeley study, also, they were rated relatively high on domineering after the age of 10, while the sister was rated as submissive. In short, although much affected by the older sibling, they are not simply assimilating the older sibling's traits. There are many symptomatic evidences of counteraction. On our own sex role measures this same boy is the least masculine at ages 11 and 19 and he is among the most anxious at both those age levels. On the Sampson and Hancock and Bragg and Allen data mentioned above, he is, along with the M1M, the most affiliative and conforming of the male dyads. On our MMPI scores this male group was the only one of these college student groups which had a deviant profile, though, in general, boys with sisters were significantly more depressive, psychopathic, and hypomanic than boys with brothers. In some more recent data, however, Leventhal (1966), sampling from a group of North Carolina college sophomores, found that although the MF2 were high on anxiety (as we had found), they also showed more interest (than MM2) in such "masculine" activities as camping, hiking, farming, horseback riding, engineering, and the study of hi-fi and stereo equipment. In addition they showed more physical strength, at least as indicated by their ability to do chinning and swimming in the physical education classes. By contrast the boys with older brothers were more interested in entrepreneurial activities such as group leadership training, the school newspaper, and occupations such as sales, business administration, and management. This latter finding was also true of the Bowling Green sample, where boys with brothers were more interested in entrepreneurial occupations than were boys with sisters. This latter data from Leventhal appear not completely consistent with the rest of the record. In Leventhal's data the boys with older sisters are being strenuously masculine. Their chin-up records and their interest in camping appear to contradict the other data in the account given above, except perhaps it can be said that even at age 6 these boys appeared to be restless under the thraldom of their older sister. Furthermore, Morris Rosenberg (1965) found these particular boys to be high in self-esteem. In order to explain the quite unusual case of this type of boy, Morris Rosenberg argues that having had only a girl, both parents are particularly pleased to give birth to the younger boy. Sears, Maccoby, and Levin (1957) reported the mother to be particularly warm to this boy if they had had several daughters already. M. Rosenberg indicates that this boy is more often cited by the older sister as a favorite with the parents. Such in fact is the self-esteem that this parent warmth fosters in him that according to Rosenberg he is relatively impervious to the usual desires for social participation, social leadership, and academic success.

Perhaps the Leventhal data (1966) can be reconciled with the rest of this record by introducing a developmental consideration along the following lines. The younger boy, because of his parents' attitudes, is by and large content with his position in the family, but being so family centered is more than ever strongly affected by the traits of the older sister, as most of the information cited above indicates, and his resistance to this is shown in symptomatic counteractive behavior. When he reaches college, however, he becomes aware of the importance of other more "masculine" considerations and makes a strenuous effort to adopt masculine traits which hitherto had not seemed so important to him. Of a confirmatory character is a study of ours in which the responses of two groups of these boys to the Gough Femininity Scale were compared. The first group were younger boys whose sisters were at college; the second group were similar younger boys who were themselves at college and whose sisters had now left home. In the first case the younger boys' scores correlated with those of their sisters and mothers. In the second case the younger boys' scores showed no such correlation, from which we may infer that once the older sister leaves the family setting, or when the boy gets to college, or both, his responses take on the more independent character that Leventhal's data (1966) seem to imply.

As a side note we would suggest that data of this sort cannot fully be understood without some underlying theory as to the nature of sex role development. Our predilection is for the view that sex role development involves the acquisition of a variety of structurally different repertoires throughout the developmental period. The modal sequence is for a boy to acquire the beginning of the affective-humanistic repertoire at his mother's knee during the first five years, the athletic-aggressive repertoire from his peers during the next ten years, and the entrepreneurial-managerial repertoire from his teachers thereafter. Analogously the girl acquires, in turn, the affective-humanistic, the nurturant-domestic, and the feminine-glamorous repertoires. Each repertoire is structurally distinct in the way in which it operates and yet is a part of a functionally continuous acquisition of modal patterns of masculinity or femininity. In particular individuals the profile will, of course, vary in the response strengths of each repertoire. While modally individuals should be capable of manifesting all the sex-role typical repertoires as the occasion requires—the male acts as father, war hero, and businessman; the female as mother, housekeeper, and attractive socialite —it is clear that many individuals are specialists in one or the other repertoire and deficient in some of the others. The research literature is partly an account of the way in which such specialities develop, particularly for males who develop aesthetic sensitivities, and females who develop entrepreneurial ones (Maccoby, 1967). In the case of the FM2, it might be argued that in childhood he is largely confined to the first stage of male repertoires and the first two stages of female repertoires by the hegemony of his sister and the warmth of the family setting, but belatedly he moves more completely to the stage-two masculine reper-

toire, long after most boys have moved to stage-three masculine repertoire.

The Berkeley Guidance data can be regarded as providing some
tentative support for this hypothesized developmental trend. This boy is
highly quarrelsome at all ages, which is consistent with the power studies
in this book (Chapter 3). With the other variables, however, the trend
is for this boy to be dependent, noncompetitive, and submissive until
about 10 years of age, after which time he moves toward more average
scores on these variables. He shifts closer to the norms during preadolescence, but does not exceed them. The Berkeley picture of anxiety
is one of very low anxiety at age 6, but of increasing anxiety until age
14, when there is again movement toward the norm.

Second-born Females of Opposite Sex (MF2)

On the recreative inventory in Chapter 3 the girl with the older
brother showed significantly more interest in athletic recreations than
the all-female dyads (F1F and FF2). When compared with F1F, FF2,
and F1M, she had the most masculine scores on Koch's ratings at age 6,
and on our self-report inventories at ages 11 and 19. She was the least
feminine and least anxious on the Minnesota Multiphasic Personality
Inventory at age 19 (Sutton-Smith & Rosenberg, 1965a). In a study of
students of that same age level, using the Strong Vocational Interest
Inventory, we also found her more interested in entrepreneurial occupations (life insurance, buyer) than the other girl dyads. It was noted
above that on the Sampson and Hancock EPPS adolescent data (1967)
she had the lowest score (for a girl dyad) on affiliation, and on the
Bragg and Allen experimental conformity data with college students
she again had the lowest conformity score for girls. In a study by
Kammeyer (1966) of siblings' attitude to female roles, he found these
same girls with older brothers to have the most masculine conception of
female traits. They tended not to agree that women are more emotional,
less aggressive, less capable of leadership, more sympathetic. In a preliminary report Altus (1966) has mentioned that more MF2 get to college than FF2, and when at college they are overrepresented among
physical education majors (Landers & Lüschen 1966). In Koch's data
on six-year-olds (1955b), these girls are said to be the most stimulated
of all the girl groups with many tomboyish qualities. Relative to the
other girl groups they were quarrelsome, tenacious, revengeful, selfish,
competitive, and confident, as well as enthusiastic, popular, and high on
leadership. In their Children's Apperception Test themes, according to
Koch (1960), they told stories involving comparisons of the relative
power of the characters in the stories. They also showed a greater expressed desire to become the opposite-sex sibling than did any other

group. In the chapter on power it was noted that they were the least powerful of all subjects vis-à-vis siblings. In the Berkeley data their scores were similar to those of their older brothers throughout. In all, this is a surprisingly unmitigated record of the older male sibling's influence on the younger girl.

Firstborn Males of Opposite Sex (M1F)

The older siblings with younger siblings of the opposite sex (M1F and F1M) tend, like the FF2 and MM2, to take extreme scoring positions. As firstborn, both should be more dependent. In fact the girl with the younger brother is more dependent, and the boy with the younger sister is not. Our favored explanation, following Koch, is that displacement by an opposite-sex younger sibling is sufficiently threatening to sex role identity at age 6 years to cause a strong counteractive identification with their own characteristic sex role. Perhaps threatened by the different sex of the younger sister, this firstborn male counteracts by becoming even more masculine, whereas the M1M, whose masculinity was not threatened by the younger sibling, could attempt to get closer to the mother by showing more "feminine" traits. By and large Koch favors the interpretation that this older sibling (M1F) prospers over the other (M1M) because his competitor (the younger sister) is simply so much less powerful, cannot keep up with him in his play with peers, and does not rival him as a male in his relationship with the mother. Koch characterizes him as aggressive, self-confident, curious, and planful.

But there is a paradox in the data. This boy (M1F) is at one extreme on two of these ratings, and near the other extreme on two of the others (Table 9.1). In the Berkeley Guidance data, he shows a slight trend toward being less domineering, less quarrelsome, less independent, but more anxious during the adolescent years. Why does the M1F seem to retain some of his infant first-born dependency responses while changing others? For the present, a speculative suggestion is that, although M1F, as well as F1M, is moving in a masculine direction during this period, the girl, being given more actual surrogate responsibilities, gets reinforcement for her executive capacities as Sampson (1962) has argued, but the M1F does not, so that his "maleness" is more in his posture than his competence.

Firstborn Females of Opposite Sex (F1M)

The F1M with the younger brother is pictured by Koch (1955a) as a highly stimulated girl. She scores the highest of all groups on most positive characteristics. She is more curious, original, enthusiastic, cheerful, ambitious, and tenacious than the other six-year-olds in these comparisons. But she is also high on such negative characteristics as jealousy, competitiveness, exhibitiveness, aggression, quarreling, and being talkative. Again, to teachers she is friendly, socially expansive, and is noted for her leadership. In her Children's Apperception Test stories, however, she demonstrates a preoccupation with mother-child relations of both a positive and negative character, and with accounts of favoritism (Koch, 1960). In her interview she expresses the view that she would like to change places with her younger male sibling, whom she sees as getting more cuddling and attention and having to be less responsible. She quarrels with him a great deal. They fight, and in his play with her he is most vigorous and, as we have seen, has power tactics to match hers. Koch interprets the girl as challenged and stimulated by her mother's additional attention to the younger brother, whom she suggests, also has the greater interest of his father. This additional attention of the father is frustrating to the girl, who increases her output of both positive (mother- and teacher-approved) characteristics and her jealousy of the younger male sibling. The girl, being older, is strong enough to react to the younger sibling with increased drive ("Thwarting within limits increases drive," Koch, 1955a, p. 29). Besides her rivalry for the younger sibling's place with the mother, she is also directly challenged by his male vigorousness. Her response is to identify more strongly with both the mothers and teachers in her environment. In Table 9.2, she is one of the most consistent in being rated at the dependent end in all studies. In the Berkeley Guidance Study she is the most submissive, most dependent, most anxious, but also the most competitive of the females throughout. Apparently she continues, as Koch describes her, as a highly feminine, yet stimulated, girl.

The Only Child

The only children have the complex parent model, the difficult task of achievement, and the explicit and verbal guidance, so they should be like the other early born in being dependent and achieving. The evidence indicates that only children are both more dependent and more achieving. They are the most self-esteeming, and the group that is chiefly responsible for the traditional view that the early born are especially eminent. Although they were branded in the case studies during the 1920's as "spoiled brats," subsequent quantitative and reliable comparisons between only children and others showed that they had no special disadvantages (Campbell, 1933; Guilford & Worcester, 1930; Hooker, 1931). Whether they, as a group, have a penchant for certain types of clinical syndromes under stress is another question, and if investigated might serve to reconcile discrepancies derived from only children as patients versus only children as normal subjects. It has been established, for example, that there are ordinal differences in the conditions under which stress occurs (Dohrenwend, 1966; Gundlach & Riess, 1967).

Unlike the firstborn, the only children are not affected by a younger sibling. The only children, unlike the firstborn, are relatively high on aggression and low on anxiety (Sears, 1951). It might well follow, therefore, that the younger sibling is one of the reasons for the lowered aggression and heightened anxiety among firstborn. Comparison of the types of influence attempts made by only children and firstborn in social situations might help to distinguish between influence attempts derived from role surrogation experiences (which only children do not have) and attempts derived from efforts to model after parents.

The most striking data on only children have to do with the sex role differences between male and female only children. Cushna's data (1966) show that mothers favor only boys to a much greater extent than only girls. There are other data to show that the only boy is more feminine than other males, and the only girl more masculine; moreover, that the deviation in these opposite-sex directions leaves them with a greater general tendency toward sex deviations consonant with these tendencies (Hooker, 1931; Heilbrun & Fromme, 1965; Rosenberg & Sutton-Smith, 1964b; Gundlach & Riess, 1967).

If this data can be taken at face value, the close relationship of only children to their parents, which is facilitative of achievement, introduces the risk of too strong an identification with the opposite-sex parent. Further evidence of this identification with the opposite-sex parent is provided by a study of Cryan (1968) with only girls. A large group of only girls were categorized as high or low on femininity on the Gough Femininity Scale, and their parents' scores on the same scale were then examined. The fathers' scores varied directly with those of the daughters, and the mothers', inversely. Furthermore, the fathers' scores correlated both with the only girls who were low in femininity and with their mothers in these groups, but did not correlate significantly with the only girls who were high in femininity or their mothers. If these correlations between father, daughter, and mother can be read as a measure of adjustment within the family, then the only-child-girl family, which is family-syntonic, is that in which both daughter and father are highly masculine and the mother is highly feminine. In the absent father data presented in this chapter, the only girls' cognitive scores were deleteriously affected by the fathers' absence, but the only boys' were not. Perhaps the simplest explanation might be that whereas the only girl loses a relationship when the father is absent, the only boy can replace one. He can become a substitute father. Whatever the interpretation, the only girls' special relationship to the father is again strongly indicated.

The Middle Born

Middle born are dealt with most extensively in Chapter 5. They seem to do more poorly in achievement (with the exception of some middle born in Rosen's data), to be more aggressive, less popular, and more role diffuse (Brock & Becker, 1965).

Whether their difficulties are due to an overall lack of interaction with the parents as compared with the others, or because they are simply heir to the structural defects of both second- and firstborn positions (without their assets) is uncertain. Like the firstborn, they may be made more rivalrous by younger infants, and like the second born, they do not occupy the power positions in the family hierarchy.

The Last Born

Most of the evidence from the youngest or late last born shows them to be most like only children, though there are exceptions (Datta, 1968). They have the assets of the firstborn (achievement, popularity), but not their handicaps (anxiety). Like firstborn, their models are either distant (because they are so much younger) or omnipresent because they have the mother to themselves, so that primary modeling norms prevail. Short-age-gap last born should be more like second born as a secondary modeling situation then prevails. There is evidence that this is the case (Miller & Zimbardo, 1966).

CONCLUSION

In conclusion, the accounts given above indicate that the M1M, MM2, MF2, and F1M follow the expectations that would be made of firstborn and second born, after the Schachter development model. If F1M and M1M are contrasted with MM2 and MF2, they should differ systematically in the predicted direction on measures of dependency, affiliation, and conformity. Contrarily, F1F and M1F, if compared with FF2 and FM2, are liable to reverse or nullify the same expectations for firstborn versus second-born differences.

A succinct way of stating the differences we have observed is to indicate that second born will acquire characteristics like those of their older siblings, and firstborn will acquire characteristics opposite in sex role character to those of their younger sibling.

The explanation of ordinal differences has departed substantially from that provided by Schachter. There seems to be little evidence for their sole infant determination. There is, however, considerable evidence for a constancy in dependency as a birth order differential response system through the preschool years for some of the ordinal positions.

Some firstborn siblings are more dependent than some later-born siblings, and for these selected groups (M1M versus MM2, and F1M versus MF2) an infant-deterministic thesis represents, in the meantime, a reasonable formulation of their developmental situation.

In an attempt to explain why some ordinal positions show considerable constancy in functioning throughout development and some show considerable change, it has been necessary to invoke a number of explanatory paradigms. Thus, the constructs of sibling modeling, sibling rivalry, and sex role identification, all of which have been advanced as normative explanations in the developmental literature, are suggested here to have relevance for particular ordinal positions. Sometimes these influences are consistent with the infant ones stressed by Schachter, sometimes they are not, and what we have in effect is a picture of development, although still incomplete, that adds to and supplements, rather than displaces, the Schachter developmental account. What this review also suggests most clearly is the need for a series of age difference studies of affiliation and conformity in which the ordinal positions are treated distinctly, and predictions are made along the lines suggested above.

conclusion

This work may be terminated by a brief discussion of three questions. What has been established empirically? What theoretical conclusions may be drawn? What practical difference does it make anyway.

EMPIRICAL

At base level, many sibling differences have been reliably established. While much of the literature makes contrasts only between the firstborn and the later born, either those contrasts or the more careful ones that take into account the position, sex, and age spacing of the siblings yield many replicated differences in sibling status. Some of the consistent differences for later born are traced in the chapters on sex role, on power, and on types of interactions (Chapters 3, 4, and 8), and for only children and firstborn siblings in the chapters on eminence, affiliation, conformity, and interaction (Chapters 5, 6, 7, and 8). The major focus of the present work upon the effects of siblings upon each other has acquired justification mainly in the studies of the later born. The sibling differences of the later born have lent themselves to interpretation almost exclusively in terms of the influence of the older sibling. Indeed, apart from the suggestion that the parents do not pay as much attention to the later born as they do to the firstborn, there has been strange lack of data on parent and later-born child relationships throughout the entire body of the data surveyed. Whether this reflects a real lack of parental influence on later born or a failure of investigators to concern themselves with this particular influence remains an empirical question. It is conceivable that the very special relationships of the parents and the firstborn have simply dominated the view of most investigators.

THEORETICAL

The data have been construed in terms of operant, modeling, and counteractive learning concepts, and the point has been stressed that the comprehension of sibling differences requires the setting of these influences into the context of family interactional structures. It was argued that it is doubtful if predictions can be made from inferred family relationships to experimental designs without allowing those designs to represent the patterns of interactions, the types of response, and the modalities of learning that are taking place in the family itself. By and large, the concepts of operant learning theory seemed to be most applicable to the development of affiliative, conforming, and achieving responses in only children and firstborn, and the concepts of modeling theory, to the development of sex role preferences and power tactics in later born. Whether these differences truly represent different sibling preferences for modes of learning or an accident derived from the variables studied is an open question. There was also some minor evidence of firstborn modeling after parents. Again, although the evidence appeared to imply that the later born modeled extensively after the elder siblings, little data were available about the way in which their behavior may have been operantly conditioned by the elder siblings. It could be expected that such a type of influence would occur.

While much of the behavior of the later born was described as a reaction (role flexibility) or counteraction (aggression) to the behavior of the elders, little was done to conceptualize these phenomena in learning terms. Presumably, the responses classed as "reactions" or "counteractions" are already a part of the later born's repertoire, so in a strict sense are not learned at all, but are selected by the subject as an appropriate response to his circumstances. It might be argued that something like a negative operant learning situation is involved, because operant learning also implies the shaping of responses rather than the acquiring of new ones. However, in this negative situation, the conditions created by the elders in the interaction are such that the later born get a greater reward by reacting in these distinctive ways (aggression, empathy), than through modeling after the elders' behavior or doing what they are told.

The bias of the child development literature of the past thirty years has been to explain personality development (hence sibling differences) in terms of early parent-child treatments. While the evidence for this position has not been impressive (Orlansky, 1949; Stevenson, 1957; Yarrow, 1964), in terms of Schachter's speculations, it has provided the only developmental proposition of influence in the existing sibling literature. Yet as the data of the previous chapter clearly show, when particular sibling positions are considered separately the data are most complex. For some of the firstborn and some of the later-born positions one could argue, as Schachter did, that the early parental treatments established response systems which remained highly consistent over time. A closer look at the data, however, seems to show that where this has occurred, it has done so with the aid and support of later developmental contingencies. Events which have transformed the pattern of development for some siblings have merely confirmed it for these apparently consistent ones. For various siblings the effect of sex role modeling after the older siblings, the rivalry with the younger siblings, or the influence of peers apparently interpenetrate with these early parent-child determinations to produce a variety of transformations (Chapter 9). The sibling data thus can be read to provide support for those, usually of behaviorist persuasion, who tend to think of human development in terms of early determinacy and constancy of effects, or in support of those organismic theorists who see the human organism as normally an open system undergoing various basic (not merely phenotypic) transformations in the course of development. There is the paradox that a concentration upon the firstborn might give an investigator the intuitive conviction that early parental actions and expectancies have a lifelong effect upon human development; whereas a focus upon the later born might give an investigator the conviction that humans are reactive to changing circumstances and hard to account for on the basis of early treatment. The notion of human development as a relatively "closed system" after the first few years has been supported by the literature of traumatic influences (with both humans and animals); the notion of human development as a relatively "open system" has received its support from the major longitudinal studies which show few such consistencies from early childhood to later growth, except in variables with an apparently genetic base. Firstborn as relatively "closed systems" may be used to provide support for closed system theorists, and later born as relatively "open systems" may be used to provide support for open system theorists.

PRACTICAL*

Sibling research as a part of social science research on human behavior, in general, is a part of the larger process of cultural differentiation that has been proceeding at a great pace throughout this century. As a result of this type of research, people are confronted with ever-increasing numbers of choice points in the course of their own development and behavior. Think, for example, of the effects of knowledge of child-rearing behavior, mediated by Spock, upon generations of parents. The behavior of any given set of parents and their children has the potentiality for becoming, in consequence, less stereotypic. For some, this is an anxiety-producing circumstance and they respond by seeking safety in older stereotypes; for others, it is a source of innovative satisfaction. Unfortunately, there is little research as yet on the way in which members of advanced cultures are managing the behavioral revolution in which they are unwittingly participating.

We should perhaps think of this sort of psychological "fission" in the same way in which we think of atomic fission, with all the implied possibilities for good and evil. The knowledge wrought by social science may be used for the greater freedom of men or for their greater control. More knowledge of others contributes to our ability to control them. On the other hand, the same knowledge of themselves permits subjects to counteract their normative dispositions, and decide not to act like, say, a typical male with a younger sister, or, alternatively, to accept these dispositions and enjoy them. When the knowledge produced by social science thus leads to novel reactions, then the social science information becomes itself a type of historical data. At least, it does if it causes change, in which case its documents become the history of the subjects' habits as they used to be. Naturally, social science keeps running and in due course understands something about the nature of these reactions and counteractions, and having created that knowledge gives rise to yet further secondary and tertiary reactions. This is what we mean by this book contributing to a process of continuing differentiation.

In more specific terms, the knowledge of how siblings characteristically react contributes to our understanding of and response to their behavior. Presumably most of us react most of the time to other people in terms of stereotypes. We are all familiar with those adults who have, say, a stereotyped view of how girls should behave and how boys should behave. These have been difficult times for such people, as the process

---

* This section was written in response to one of our more activist students, who challenged the practical worth of our enterprise. We were grateful for the stimulus, and suggested that his questioning was itself an example of the differentiation we discuss above.

of sex role differentiation has been proceeding at a great pace (a topic which we take up in a companion volume). It is probably more desirable for such persons to have a whole drawerful of stereotypes or categories to use than to operate on only one or two. But even for persons who have sufficient compassion and passivity in their nature that their reponses are not unduly biased by preexisting stereotypes, there is always an initial period where some responses must be made to the individually different person. The present volume may add a fragment to their sensitivity.

And finally, of even greater practical importance, everyone is a member of the sibling game. Most people want to have the opportunity to think about their own lives, about those long "naked" years with their siblings, and to make decisions as to whether, after all, they will be modally an F1F and FF2, or even an MM2 and a MMFM4M.

References

Abernathy, E. M. Data on personality and family position. *Journal of Psychology,* 1940, **10,** 303–307.

Ackerman, N. W. *Psychodynamics of family life.* New York: Basic Books, 1958.

Adams, R. L. Personality and behavioral differences among children of various birth positions. *Dissertation Abstracts,* 1967, Vol. 28.

Adler, A. Characteristics of the 1st, 2nd, and 3rd child. *Children,* 1928, **3, 14**(Issue 5).

Adler, A. *Understanding human nature.* New York: Premier Books (Fawcett Publications), 1959.

Alexander, C. N. Ordinal position and sociometric status. *Sociometry,* 1967, **29,** 41–51.

Altus, W. D. Birth order, intelligence and adjustment. *Psychological Reports,* 1959, **5,** 502.

Altus, W. D. Sibling order and scholastic aptitude. *American Psychologist,* 1962, **17,** 304.

Altus, W. D. Birth order, aptitude and the Gottschaldt Test. Address, American Psychological Association Meeting, Chicago, September 1965.

Altus, W. D. Birth order and its sequelae. *Science,* 1966, **151,** 44–49. (a)

162   The Sibling

Altus, W. D. *Birth order and the omnibus personality inventory. Proceedings,* American Psychological Association Convention, 1966, 279–280. (b)

Altus, W. D. *Birth order and the choice of college major. Proceedings,* American Psychological Association Convention, 1967, 287–288. (a)

Altus, W. D. Birth order, femininity and sex of sibling. Paper presented at the American Psychological Association, Washington, D.C., September 1967. (b)

Amir, Y., Sharan, S., & Kovarsky, Y. Birth order, family structure and avoidance behavior. *Journal of Personality and Social Psychology,* 1968, Vol. 10, 271–278.

Ansbacher, H. L., & Ansbacher, R. R. *The individual psychology of Alfred Adler.* New York: Harper & Row, Harper Torchbooks (TB 1154), 1956.

Armilla, J. Birth order research: Participation of Peace Corp volunteers. *Psychological Reports,* 1966, 18, 56–58.

Arrowood, J., & Amoroso, D. M. Social comparison and ordinal position. *Journal of Personality and Social Psychology,* 1965, 2, 101–104.

Atkinson, J. W., Bastian, J. R., Earl, R. W., & Litivin, G. H. The achievement motive, goal setting, and probability preferences. *Journal of Abnormal and Social Psychology,* 1960, 60, 27–36.

Atkinson, J. W., & Litivin, G. H. Achievement motive and test anxiety conceived as motive to approach success and motive to avoid failure. *Journal of Abnormal and Social Psychology,* 1960, 60, 52–63.

Bakan, D. The relationship between alcoholism and birth rank. *Quarterly Journal for the Study of Alcoholism,* 1949, 10, 434–440.

Baker, H. J., Decker, F. J., & Hill, A. S. A study of juvenile theft. *Journal of Educational Research,* 1929, 20, 81–87.

Baldwin, A. L. Differences in parent behavior toward three- and nine-year-old children. *Journal of Personality,* 1946, 15, 143–165.

Baldwin, A. L., Kalhorn, J., & Breese, F. H. The appraisal of parent behavior. *Psychological Monographs,* 1949, 63, 1–85.

Bandura, A., & Walters, R. H. *Social learning and personality development.* New York: Holt, Rinehart and Winston, 1963.

Barron, F. *Creativity and psychological health.* New York: Van Nostrand, 1963.

Barry, H., & Barry, H., Jr. Birth order, family size, and schizophrenia. *Archives of General Psychiatry,* 1967, Vol. 17, 435–440.

Bartlett, E. W., & Smith, C. P. Child-rearing practices, birth order, and the development of achievement-related motives. In E. D. Evans (Ed.), *Children: Readings in Behavior and Development.* New York: Holt, Rinehart and Winston, 1968, 58–68.

Bauer, M. L., & Erlich, H. J. Sibling sex distribution and psychiatric status. *Psychological Reports,* 1966, 18, 365–366.

Bayer, A. E. Birth order and college attendance. *Journal of Marriage and Family Living,* 1966, 28, 480–484.

Bayer, A. E. Birth order and attainment of the doctorate: A test of an economic hypothesis. *American Journal of Sociology,* 1967, 72, 540–550.

Bayer, A. E., & Folger, J. K. The current state of birth order research. *International Journal of Psychiatry*, 1967, **3**, 37–39.

Bayley, N. Comparisons of mental and motor test scores for ages 1–15 months by sex, birth order, race, geographical location, and education of parents. *Child Development*, 1965, **36**, 379–411.

Bayley, N., & Schaefer, E. S. Relationships between socioeconomic variables and the behavior of mothers toward young children. *Journal of Genetic Psychology*, 1960, **96**, 61–67.

Becker, C. The effects of group therapy upon sibling rivalry. *Smith College Studies in Social Work*, 1945, **16**, 131–132.

Becker, G. Affiliate perception and the arousal of the participation-affiliation motive. *Perceptual and Motor Skills*, 1967, **24**, 991–997.

Becker, S. W., & Carroll, J. Ordinal position and conformity. *Journal of Abnormal and Social Psychology*, 1962, **65**, 129–131.

Becker, S. W., & Lerner, M. J. Conformity as a function of birth order and types of group pressure. *American Psychologist*, 1963, **18**, 402.

Becker, S. W., Lerner, M. J., & Carroll, J. Conformity as a function of birth order and type of group pressure; A verification. *Journal of Personality and Social Psychology*, 1966, Vol. 3, No. 2, 242–244.

Becker, W. C., Peterson, D. R., Hellmer, L. A., Shoemaker, D., & Quay, H. C. Factors in parental behavior and personality as related to problem behavior in children. *Journal of Consulting Psychologists*, 1959, **23**, 107–118.

Bell, N. W., & Cragel, E. F. *A modern introduction to the family*. Glencoe, Ill.: Free Press, 1960.

Bell, R. Q. A reinterpretation of the direction of effects in studies of socialization. *Psychological Review*, 1968, Vol. 75, 81–95.

Bellrose, D. *Behavior problems of children*. Master's thesis. Smith College School for Social Work, 1927.

Bender, I. E. Ascendance-submission in relation to certain other factors in personality. *Journal of Abnormal and Social Psychology*, 1928, **23**, 137–143.

Bene, E., & Anthony, J. *Manual for the family relations test*. London: National Foundation for Educational Research, 1957.

Bennett, I. *Delinquent and neurotic children*. New York: Basic Books, 1960.

Berman, H. H. Order of birth in manic-depressive reactions. *Psychiatric Quarterly*, 1933, **7**, 430–435.

Bernard, J. The adjustments of married mates. In H. T. Christensen (Ed.), *Handbook of marriage and the family*. Chicago: Rand McNally, 1964.

Berne, E. *Games people play*. New York: Grove Press, 1964.

Bieber, I., Dain, H. J., Dince, P. R., Drellich, M. G., Grand, H. G., Gundlach, R. H., Kremer, M. W., Rifkin, A. H., Wilbur, C. B., & Bieber, T. B. *Homosexuality: A psychoanalytic study*. New York: Basic Books, 1962.

Blatz, W. E., & Bott, E. A. Studies in mental hygiene of children: I: Behavior of public school children—A description of method. *Journal of Genetic Psychology*, 1927, **34**, 552–582.

Blum, G. S. A study of the psychoanalytic theory of psychosexual development. *Genetic Psychology Monographs,* 1949, **39,** 3–102.

Bohannon, E. W. The only child in a family. *Pedagogical Seminary,* 1898, **5,** 474–496.

Bossard, J. H. S. Family modes of expression. *American Sociological Review,* 1945, **10,** 226–237.

Bossard, J. H. S. A spatial index for family interaction. *American Sociological Review,* 1951, **16,** 243–246.

Bossard, J. H. S. *Parent and child: Studies in family behavior.* Philadelphia: University of Pennsylvania Press, 1953.

Bossard, J. H. S., & Boll, E. Security in the large family. *Mental Hygiene,* 1954, **38,** 529–544.

Bossard, J. H. S., & Boll, E. Personality roles in the large family. *Child Development,* 1955, **26,** 71–78.

Bossard, J. H. S., & Boll, E. Adjustment of siblings in large families. *American Journal of Psychiatry,* 1956, **112,** 889–892. (a)

Bossard, J. H. S., & Boll, E. *The large family system.* Philadelphia: University of Pennsylvania Press, 1956. (b)

Bossard, J. H. S., & Boll, E. *The sociology of child development.* New York: Harper & Row, 1960.

Bossard, J. H. S., & Sanger, M. The large family. *American Sociological Review,* 1952, **17,** 3–9.

Boszormewyi-Nagy, I., & Framo, J. L. (Eds.) *Intensive family therapy: Theoretical and practical aspects.* New York: Harper & Row, 1965.

Bower, P., & London, P. Developmental correlates of role-playing ability. *Child Development,* 1965, **30,** 499–508.

Bowerman, C. E., & Day, B. L. A test of the theory of complementary needs as applied to couples during courtship. *American Sociological Review,* 1956, **21,** 602–605.

Bowerman, C. E., & Elder, G. H. Variations in adolescent perception of family power structure. *American Sociological Review,* 1964, **29,** 551–567.

Bragg, B. W. E., & Allen, V. L. Ordinal position and conformity. Paper presented at the American Psychological Association, Washington, D.C. September, 1967.

Brim, O. G. Family structure and sex role learning by children: A further analysis of Helen Koch's data. *Sociometry,* 1958, **21,** 1–16.

Brim, O. G. Some basic research problems in parent education with implications from the field of child development. *Monographs of the Society for Research in Child Development,* 24, 1959, 51–68.

Brittain, C. V. Age and sex of siblings and conformity towards parents versus peers in adolescence. *Child Development,* 1966, **37,** 709–714.

Brock, T. C., & Becker, G. Birth order and subject recruitment. *Journal of Social Psychology,* 1965, **65,** 63–66.

Bronfenbrenner, U. Some familial antecedents of responsibility and leadership in adolescents. In P. Petrullo and B. Bass (Eds.), *Leadership and Interpersonal Behavior.* New York: Holt, Rinehart and Winston, 1961. Pp. 239–271.

Bronson, W. C. Central orientations: A study of behavior organization from childhood to adolescence. *Child Development,* 1966, Vol. 37, No. 1, 125–155.

Broverman, D. M., & Klaiber, E. L. The automatization cognitive style and birth phenomena: Maternal age and birth order. Manuscript, Psychology Department, Worcester State Hospital, Worcester, Mass. 1968.

Brunswik, E. *Perception and the representative design of psychological experiments.* Berkeley: University of California Press, 1956.

Burchinaz, L. G. The premarital dyad and love involvement. In H. T. Christensen (Ed.), *Handbook of the family.* Chicago: Rand McNally, 1964.

Burt, C. *The young delinquent.* New York: Appleton, 1925.

Burton, D. Birth order and intelligence. *Journal of Social Psychology,* 1968, Vol. 76, 199–206.

Byrdy, H. L., & Uhlenhuth, E. H. Ordinal sibling position and motivation toward psychiatric treatment. *The Journal of Nervous and Mental Disease,* 1962, **135,** 265–267.

Campbell, A. A. A study of the personality adjustments of only and intermediate children. *Journal of Genetic Psychology,* 1933, **43,** 197–206.

Capra, P. C., & Dittes, J. E. Birth order as a selective factor among volunteer subjects. *Journal of Abnormal and Social Psychology,* 1962, **64,** 302.

Carlsmith, L. Effect of early father absence on scholastic aptitude. *Harvard Educational Review,* 1964, **34,** 3–21.

Carman, A. Pain and strength measurements of 1,507 school children in Saginaw, Michigan. *American Journal of Psychology,* 1899, **10,** 392–398.

Carrigan, W. C., & Julian, J. W. Sex and birth order differences in conformity as a function of need-affiliation arousal. *Journal of Personality and Social Psychology,* 1966, **3,** 479–483.

Caudill, W. Sibling rank and social background among Japanese psychiatric patients. In R. Dore (Ed.), *Social change in modern Japan.* Princeton, N.J.: Princeton University Press, 1966.

Chambers, J. Relating personality and biographical factors to scientific creativity. *Psychological Monographs,* 78(No. 584), 1964.

Chen, E., & Cobb, S. Family structure in relation to health and disease. *Journal of Chronic Disorders,* 1960, **12,** 544–567.

Chittenden, E. A., Foan, W., Zweil, D. M., & Smith, J. R. School achievement of first and second born children. *Child Development,* 1968, Vol. 39, 1123–1129.

Chopra, S. Family size and sibling position as related to intelligence test scores and academic achievement. *Journal of Social Psychology,* 1966, **70,** 133–137.

Christensen, H. T. (Ed.) *Handbook of marriage and the family.* Chicago: Rand McNally, 1964.

Christie, R. "Machiavellianism." Address delivered at the Social Psychology Seminar, Teachers College, Columbia University, 1967.

Clausen, J. A. Family structure, socialization and personality. In L. W. Hoffman, and M. L. Hoffman (Eds.), *Review of Child Development Research,* New York: Russell Sage Foundation, 1966. Pp. 1–54.

Cleland, C. C., Seitz, S., & Patton, W. F. Birth order and ruralism as potential determinants of attendant tenure. *American Journal of Mental Deficiency,* 1967, Vol. 72, No. 3, 428–434.

Cobb, S., & French, J. R. P. Birth order among medical students. *Journal of the American Medical Association,* 1966, **195,** 172–173.

Cohen, F. Psychological characteristics of the second child as compared with the first. *Indian Journal of Psychology,* 1951, **26,** 79–84.

Collard, Roberta. Social and play responses of first born and later born infants in an unfamiliar situation. *Child Development,* 1968, **39**(1), 325–334.

Conners, C. K. Birth order and needs for affiliation. *Journal of Personality,* 1963, **31,** 408–416.

Craik, K. A comparison of the personal histories of a group of creative architects, a group of architects associated with creative architects, and a representative group of architects. Mimeograph, University of California, 1961.

Crump, E. P., Horton, C. P., Masuoka, J., & Ryan, D. Growth and development. I. Relation of birth weight in Negro infants to sex, maternal age, position, prenatal care, and socioeconomic status. *Journal of Pediatrics,* 1957, **51,** 678–689.

Cryan, J. Ordinal position and masculinity-femininity: Their influence on the female personality. Master's Thesis, Bowling Green State University, 1968.

Cummings, E., & Schneider, D. M. Sibling solidarity: A property of American kinship. *American Anthropologist,* 1961, **63,** 498–507.

Cushna, B. Agency and birth order differences in very early childhood. Paper presented at the meeting of the American Psychological Association, New York, September 1966.

Cushna, B. Birth order and verbal mastery. Paper presented at the American Psychological Association, San Francisco, September 1968.

Cushna, B., Greene, M., & Snider, B. C. F. First born and last born children in a child development clinic. *Journal of Individual Psychology,* 1964, **20,** 179–182.

Dager, E. Z. Socialization and personality development in the child. In H. T. Christensen (Ed.), *Handbook of marriage and the family.* Chicago: Rand McNally, 1964.

Damrin, D. E. Family size and sibling age, sex, and position as related to certain aspects of adjustment. *Journal of Social Psychology,* 1949, **29,** 93–102.

Darley, J. M. Fear and social comparison as determinants of conformity behavior. *Journal of Personality and Social Psychology,* 1966, 4, 73–78.

Darley, J. M., & Aronson, E. Self evaluation vs. direct anxiety reduction as determinants of the fear-affiliation relationship. *Journal of Experimental Social Psychology,* 1966(Supp. 1), 66–79.

Datta, L. E. Birth order and early scientific attainment. *Perceptual and Motor Skills,* 1967, **24,** 157–158.

Datta, L. E. Birth order and potential scientific creativity. *Sociometry,* 1968, Vol. 31, No. 1, 76–88.

Davis, A. American status systems and the socialization of the child. In C. Kluckhohn and H. A. Murray (Eds.), *Personality in nature, society and culture.* New York: Knopf, 1961. Pp. 363–375.

Dean, D. A. The relation of ordinal position to personality in young children. Master's Thesis, State University of Iowa, 1947.

De Lint, Jan E. E. Alcoholism, birth order and socializing agents. *Journal of Abnormal and Social Psychology,* 1964, **69**(4), 457–458.

Dember, W. M. Birth order and need affiliation. *American Psychologist,* 1963, **18**, 356.

Dimond, R., & Munz, C. Ordinal position of birth and self disclosure in high school students. *Psychological Reports,* 1968, Vol. 21, 829–833.

Dittes, J. E. Effect of changes in self-esteem upon impulsiveness and deliberation in making judgments. *Journal of Abnormal and Social Psychology,* 1959, **58**, 348–356.

Dittes, J. E. Birth order and vulnerability to differences in acceptance. *American Psychologist,* 1961, **16**, 358.

Dittes, J. E., & Capra, P. C. Affiliation: Comparability or compatibility. *American Psychologist,* 1962, **17**, 329.

Dohrenwend, B. S., & Dohrenwend, B. P. Stress situations, birth order and psychological symptoms. *Journal of Abnormal Psychology,* 1966, **17**, 215–223.

Dohrenwend, B. S., Feldstein, S., Plosky, J., & Schmeidler, G. R. Factors interacting with birth order in self selection among volunteer subjects. *Journal of Social Psychology,* 1967, **72**, 125–128.

Eisenman, R. Birth order, aesthetic preference, and volunteering for an electric shock experiment. *Psychonomic Science,* 1965, **3**, 151–152.

Eisenman, R. Birth order, anxiety, and verbalizations in group psychotherapy. *Journal of Consulting Psychology,* 1966, **30**, 521–526.

Eisenman, R. Birth order, insolence, socialization, intelligence, and complexity-simplicity differences. *Journal of General Psychology,* 1968, **78**, 61–64. (a)

Eisenman, R. Personality and demography in complexity-simplicity. *Journal of Consulting and Clinical Psychology,* 1968, Vol. 32, No. 2, 140–143.

Eisenman R., & Platt, J. J. Birth order and sex differences in academic achievement and internal-external control. *Journal of General Psychology,* 1968, Vol. 78, 279–285.

Elder, G. H. *Adolescent achievement and mobility aspirations.* Chapel Hill, N.C.: Institute for Research in Social Science, 1962.

Ellis, H. *A study of British genius.* Boston: Houghton Mifflin, 1926.

Erikson, E. H. *Childhood and Society.* New York, W. W. Norton, 1950.

Erlich, D. Determinants of verbal commonality and influencibility. Ph.D. Thesis, University of Minnesota, 1958.

Exner, J. Personal communication on preliminary findings in an E.A.P. report. Peace Corps Office, Washington, D.C., 1969.

Farber, B., & Jenne, W. C. Family organization and parent-child communications: Parents and siblings of a retarded child. *Monographs of the Society for Research in Child Development,* 1963, **28**, 1–78.

Fauls, L. B., & Smith, W. D. Sex role learning in five-year-olds. *Journal of Genetic Psychology*, 1956, **89**, 105–117.

Fenton, N. The only child. *Journal of Genetic Psychology*, 1927, **34**, 45–71.

Ferguson, E. D. The effect of sibling competition and alliance on level of aspiration, expectation, and performance. *Journal of Abnormal and Social Psychology*, 1958, Vol. 56, No. 2, 213–222.

Festinger, L. Informal social communications. *Psychological Review*, 1950, **57**, 271–282.

Festinger, L. Motivations leading to social behavior. In M. R. Jones (Ed.), *Nebraska symposium on motivation*. Lincoln: University of Nebraska Press, 1954. Pp. 191–219.

Finneran, M. P. Dependency and self-concept as functions of acceptance-rejection by others. *American Psychologist*, 1958, **13**, 332.

Fischer, E. H., Wells, C. F., & Cohen, S. L. Birth order and expressed interest in becoming a college professor. *Journal of Counseling Psychology*, 1968, Vol. 15, 111–116.

Foster, S. A study of the personality make-up and social setting of fifty jealous children. *Mental Hygiene*, 1927, **11**, 53–77.

Frank, G. H. The role of the family in the development of psychopathology. *Psychological Bulletin*, 1965, **64**, 191–205.

Freedman, D. S., Freedman, R., & Whelpton, P. K. Size of family and preference for children of each sex. *American Journal of Sociology*, 1960, **66**, 141–164.

French, E. G. Development of a measure of complex motivation: In J. W. Atkinson (Ed.), *Motives in fantasy, action, and society*. Princeton, N.J.: Van Nostrand, 1958. Pp. 242–248.

French, J. R. P., & Raven, B. The bases of social power. In D. Cartwright (Ed.), *Studies in social power*. Ann Arbor: University of Michigan Press, 1959.

Galton, F. *English men of science: Their nature and nurture*. London: Mac-Millan 1874.

Gerard, H. B., & Rabbie, J. M. Fear and social comparison. *Journal of Abnormal and Social Psychology*, 1961, **62**, 586–592.

Gewirtz, J. L. *Succorance in young children*. Ph.D. Thesis, State University of Iowa, Iowa City, Iowa, 1948.

Gewirtz, J. L., & Gewirtz, H. B. Stimulus conditions, infant behaviors, and social learning in four Israeli child-rearing environments: A preliminary report illustrating differences in environment and behavior between the 'only' and the 'youngest' child. In B. M. Foss (Ed.), *Determinants of infant behavior III*. London: Methuen (New York: Wiley), 1965. Pp. 161–184.

Gilmore, J. B., & Zigler, E. Birth order and social reinforcer effectiveness with children. *Child Development*, 1964, **35**, 193–200.

Glass, D. C., Brim, O. G., & Neulinger, J. Birth order, sex of sibling and achievement orientation. Report of Russell Sage Foundation, New York, 1969.

Glass, D. D., Horowitz, W., Firestone, I., & Grinker, J. Birth order and reactions to frustration. *Journal of Abnormal and Social Psychology*, 1963, **66**, 192–194.

Goffman, I. *Encounters: Two studies in the sociology of interaction.* Indianapolis: Bobbs-Merrill, 1961.

Gollob, H. F., & Dittes, J. E. Different effects of manipulated self-esteem on persuasibility depending on the threat and complexity of the communication. Paper presented at the American Psychological Association, 1963.

Goodenough, F. L., & Leahy, A. M. The effect of certain family relationships upon the development of personality. *Journal of Genetic Psychology,* 1927, **34**, 45–72.

Gordon, B. F. Influence and social comparison as motives for affiliation. *Journal of Experimental Social Psychology*, 1966, Supp. 1, 55–65.

Gordon, K., & Gordon, R. D. Birth order, achievement, and blood chemistry levels among college nursing students. *Nursing Research,* 1967, Vol. 16, 234–236.

Gough, H. Identifying psychological femininity. *Educational and Psychological Measurement,* 1952, **12**, 427–439.

Granville-Grossman, K. L. Birth order and schizophrenia. *British Journal of Psychiatry,* 1966, **112**, 1119–1126.

Green, C. E., Eastman, M. E., & Adams, S. T. Birth order, family size and extra-sensory perception. *British Journal of Social and Clinical Psychology,* 1966, **5**, 150–152.

Greenberg, H., Mayer, D., Guerena, R., Pislowski, D., & Lashen, M. Order of birth as a determinant of personality and attitudinal characteristics. *Journal of Social Psychology,* 1963, **60**, 221–230.

Greenberg, M. S. Role playing: An alternative to deception. *Journal of Personality and Social Psychology,* 1967, **7**, 152–157.

Greenberg, R., & White, C. The sexes of consecutive sibs in human sibships. *Human Biology,* 1967, Vol. 39, 374–404.

Greenfield, N. S. The relationship between recalled forms of childhood discipline and psychopathology. *Journal of Consulting Psychology,* 1959, **23**, 139–142.

Gregory, I. An analysis of familial data on psychiatric patients: Parental age, family size, birth order, and ordinal position. *British Journal of Preventive Social Medicine,* 1958, **12**, 42–59.

Griffiths, C. H. The influence of family on school marks. *School and society,* 1926, **24**, 713–716.

Griffiths, J. Identification, ordinal position, and kind of model. Manuscript, Bowling Green State University, 1966.

Grosz, H. J., & Miller, I. Sibling patterns in schizophrenics. *Science,* **128**, 1958, p. 30.

Guilford, R. B., & Worcester, D. A. A comparative study of the only and non-only children. *Journal of Genetic Psychology,* 1930, **38**, 411–426.

Gundlach, R. H., & Riess, B. F. Birth order and sex of siblings in a sample of lesbians and non-lesbians. *Psychological Reports,* 1967, **20**, 61–62.

Gundlach, R. H., & Riess, B. F. Characteristics of middle-class female homosexuals. Paper presented at the American Psychological Association, San Francisco, September 1968.

Haeberle, A. W. Interactions of sex, birth order, and dependency with behavior problems and symptoms in emotionally disturbed preschool children. Paper at Eastern Psychological Association, Philadelphia, 1958.

Hakmiller, K. L. Need for self-evaluation, perceived similarity and comparison choice. *Journal of Experimental Social Psychology*, 1966, **1**, 49–54.

Hall, C. S., & Lindzey, G. *Theories of Personality*. New York: Wiley, 1966.

Hall, E., & Barger, B. Attitudinal structures of older and younger siblings. *Journal of Individual Psychology*, 1964, **20**, 59–68.

Hall, R. I., & Willerman, B. The educational influence of dormitory roommates. *Sociometry*, 1963, **26**, 294–318.

Hamilton, M. L. Affiliative behavior as a function of approach and avoidance affiliation motives, opinion evaluation and birth order. *Journal of Social Psychology*, 1967, **72**, 61–70.

Handel, G. Psychological study of whole families. *Psychological Bulletin*, 1965, **63**, 19–41.

Harrington, C. *Sex role deviancy in adolescents*. New York: Teachers College Press, Columbia University, 1969.

Harris, I. D. *The promised seed: A comparative study of eminent first and later sons*. Glencoe, Ill.: Free Press, 1964.

Harris, I. D., & Howard, K. T. Birth order influences on moral responsibility. *Institute for Juvenile Research, Department of Mental Health Research Report* (Chicago, Ill.), 1966, **3**.

Haven, H. J. The effect of ordinal position, family size and sex on cognitive complexity. M.A. Thesis, Bowling Green State University Library, 1967.

Hawkes, R., Burchinal, L., & Gardner, B. Size of family and adjustment of children. *Journal of Marriage and Family Living*, 1958, **20**, 65–68.

Heilbrun, A. E., & Fromme, D. K. Parental identification of late adolescent and level of adjustment: The importance of parental model attributes, ordinal position and sex of child. *Journal of Genetic Psychology*, 1965, **107**, 49–59.

Helmreich, R. L. Prolonged stress in Sealab II: A field study of individual and group reactions. Ph.D. Thesis, Yale University, 1966.

Helmreich, R. L. Birth order effects. *Naval Research Reviews*. Office of Naval Research, Washington, D.C., 1968, 1–6.

Helmreich, R. L., & Collins, B. E. Situational determinants of affiliative preference under stress. *Journal of Personality and Social Psychology*, 1967, **6**, 79–85.

Helson, R. Effects of sibling characteristics and parental values on creative interest and achievement. *Journal of Personality*, 1968, **36**, 589–607.

Henry, A. F. Sibling structure and perception of disciplinary roles of parents. *Sociometry*, 1957, **20**, 67–74.

Henry, J., & Henry, Z. Doll play of Pilaga Indian children. In C. Kluckhohn and H. A. Murray, *Personality in nature, society and culture*. New York: Knopf, 1961. Pp. 292–307.

Henry, W. Identity and diffusion in professional actors. Paper delivered to American Psychological Association, Chicago, September 1965.

Herskovits, M., & Herskovits, F. Sibling rivalry, the oedipus complex and myth. *Journal of American Folklore*, 1958, **71**, 1–15.

Hess, R. D., & Handel, G. *Family worlds, A psychosocial approach to family life*. Chicago: University of Chicago Press, 1959.

Hess, R. D., Sims, J. H., & Henry, W. F. Identity diffusion and the social role of the actor. Paper delivered at the American Psychological Association, St. Louis, September 1962.

Hilton, I. Differences in the behavior of mothers toward first and later born children. *Journal of Personality and Social Psychology*, 1967, **7**, 282–290.

Hilton, I. The dependent first born and how he grew. Paper presented at the American Psychological Association, San Francisco, September 1968.

Hooker, H. F. The study of the only child at school. *Journal of Genetic Psychology*, 1931, **39**, 122–126.

Howells, W. W. Birth order and body size. *American Journal of Physical Anthropology*, 1948, **6**, 449–460.

Hubert, M. A. G. Age, ordinal position and sex related to mother-child behavior. *Journal of Home Economics*, 1957, **49**, 208–223.

Jacoby, J. Birth rank and pre-experimental anxiety. *Journal of Social Psychology*, 1968, **76**, 9–11.

Johnson, M. Sex role learning in the nuclear family. *Child Development*, 1963, **34**, 319–333.

Jones, H. E. Order of birth in relation to the development of the child. In C. Murchison (Ed.), *A handbook of child psychology*. Worcester, Mass.: Clark University Press, 1933.

Jones, H. E. The environment and mental development. In L. Carmichael (Ed.), *Manual of child psychology*. New York: Wiley, 1954.

Jones, H. E., & Hsiao, H. H. A preliminary study of intelligence as a function of birth order. *Journal of Genetic Psychology*, 1928, **35**, 428–433.

Kagan, J., & Moss, H. A. The stability of passive and dependent behavior from childhood through adulthood. *Child Development*, 1960, **31**, 577–591.

Kagan, J., & Moss, H. A. *Birth to maturity: A study in psychological development*. New York: Wiley, 1962.

Kammeyer, K. Birth order and the feminine sex role among college women. *American Sociological Review*, 1966, **31**, 508–515.

Kammeyer, K. Birth order as a research variable. *Social Forces*, 1967, **46**, 71–80. (a)

Kammeyer, K. Sibling position and the feminine role. *Journal of Marriage and the Family*, 1967, **29**, 494–499. (b)

Kayton, L., & Borge, G. S. Birth order and the obsessive-compulsive character. *Archives of General Psychiatry*, 1967, **17**(16), 751–754.

Kemper, T. D. Mate selection and marital satisfaction according to sibling type of husband and wife. *Journal of Marriage and Family Living*, 1966, **28**, 346–349.

Klaus, R. A., & Gray, S. W. The early training project for disadvantaged children: A report after five years. *Monographs of the Society for Research in Child Development,* 1968, **33,** No. 4.

Koch, H. L. The relation of "primary mental abilities" in five- and six-year-olds to sex of child and characteristics of his sibling. *Child Development,* 1954, **25,** 209–223.

Koch, H. L. The relation of certain family constellation characteristics and the attitudes of children toward adults. *Child Development,* 1955, **26,** 13–40. (a)

Koch, H. L. Some personality correlates of sex, sibling position, and sex of sibling among five and six year old children. *Genetic Psychological Monographs,* 1955, **52,** 3–50. (b)

Koch, H. L. Sibling influence on children's speech. *Journal of Speech Disorders,* 1956, **21,** 322–328. (a)

Koch, H. L. Children's work attitudes and sibling characteristics. *Child Development,* 1956, **27,** 289–310. (b).

Koch, H. L. Some emotional attitudes of the young child in relation to characteristics of his sibling. *Child Development,* 1956, **27,** 393–426. (c)

Koch, H. L. Sissiness and tomboyishness in relation to sibling characteristics. *Journal of Genetic Psychology,* 1956, **88,** 231–244. (d)

Koch, H. L. Attitudes of young children toward their peers as related to certain characteristics of their siblings. *Psychological Monographs,* 1956, **70** (Whole No. 425). (e)

Koch, H. L. The relation in young children between characteristics of their playmates and certain attributes of their siblings. *Child Development,* 1957, **28,** 175–202.

Koch, H. L. The influence of siblings on the personality development of younger boys. *Journal of Psychology and Psychotherapy,* 1958, **5,** 211–225.

Koch, H. L. The relation of certain formal attributes of siblings to attitudes held toward each other and toward their parents. *Monographs of the Society for Research in Child Development,* 1960, **25,** 1–124.

Koch, H. L. *Twins and Twin Relations.* Chicago: Univ. of Chicago Press, 1966.

Konig, K. *Brothers and sisters: A study in child psychology.* New York: St. George Books, Blauvelt, 1963.

Krebs, A. M. The determinants of conformity: Age of independence training and achievement. *Journal of Abnormal and Social Psychology,* 1958, **56,** 130–131.

Krinsky, S. The relationship among birth order, dimensions of independence-dependence, and choice of a scientific career. Unpublished doctoral dissertation, Harvard University, 1963.

Krout, M. H. Typical behavior patterns in 26 ordinal positions. *Journal of Genetic Psychology,* 1939, **54,** 3–29.

Landers, D. M., & Lüschen, G. Sibling-sex-status and ordinal position effects on the sport participation of females. Paper presented at the International Congress of Sport Psychology, Washington, D.C., October 1966.

Landy, F. S., Rosenberg, B. G., & Sutton-Smith, B. The effect of limited father absence on cognitive development. *Child Development*, 1969, **40**, 941–944.

Lansky, L. The family structure also affects the model: Sex-role identification in parents of preschool children. *Merrill-Palmer Quarterly*, 1964, **10**, 39–50.

Lasko, J. K. Parent behavior towards first and second children. *Genetic Psychological Monographs*, 1954, **49**, 96–137.

Latane, B. Studies in social comparison. *Journal of Experimental Social Psychology*, 1966, **1**, 1–5.

Latane, B., Eckman, J., & Joy, V. Shared stress and interpersonal attraction. *Journal of Experimental Social Psychology*, 1966, **1**, 80–94.

Lees, J. P. The social nobility of a group of eldest-born and intermediate adult males. *British Journal of Psychology*, 1952, **43**, 210–221.

Lessing, E., & Oberlander, M. *Ordinal position and childhood psychopathology as evaluated from four perspectives.* Proceedings, 75th Annual Convention of the American Psychological Association, San Francisco, September 1967, 179–180.

Lester, D. Sibling position and suicidal behavior. *Journal of Individual Psychology*, 1966, **22**, 204–207.

Leventhal, G. S. Sex of sibling as a predictor of personality characteristics. *American Psychologist*, 1965, **20**, 783.

Leventhal, G. S. Sex of sibling, birth order and behavior of male college students from two-child families. Manuscript, Psychology Department, North Carolina State University, 1966.

Levinson, P. A study of the relationship between sibling position and reading ability. Unpublished doctoral dissertation, University of Pennsylvania, 1963.

Levi-Strauss, C. *Structural anthropology.* New York: Basic Books, 1963.

Levy, D. M. Maternal over-protection and rejection. *Archives of Neurology and Psychiatry*, 1931, **25**, 886–889.

Levy, D. M. Sibling rivalry. *American Orthopsychiatric Association, Research Monograph #2*, 1937.

Levy, D. M. Sibling rivalry studies in children of primitive groups. *American Journal of Orthopsychiatry*, 1939, **9**, 205–215.

Lewis, M. A. The meaning of a response, or why researchers in infant behavior should be oriental metaphysicians. *Merrill-Palmer Quarterly*, 1967, **13**, 7–18.

Lowe, C. R., & Gibson, J. R. Weight at third birthday related to birthweight, duration of gestation, and birth order. *British Journal of Preventive Social Medicine*, 1953, **7**, 78–82.

Lunneborg, P. Birth order, aptitude and achievement. *Journal of Consulting and Clinical Psychology*, 1968, Vol. 32, No. 1, 101.

Lu Yi-Chang. Predicting roles in images. *American Journal of Sociology*, 1952, **58**, 51–55.

Maccoby, E. (Ed.) *The development of sex differences.* London: Tavistock, 1967.

MacDonald, A. P., Jr. Birth-order effects in marriage and parenthood: Affiliation and socialization. *Journal of Marriage and the Family,* 1967, **29,** 656–661.

MacDonald, A. P., Jr. Manifestations of differential levels of socialization by birth order. *Developmental Psychology,* 1969, **1,** 485–492a.

MacDonald, A. P., Jr. Birth order and religious affiliation. *Developmental Psychology,* 1969, **1,** 628b.

MacFarlane, J. W. Studies in child guidance. *Monograph of the Society for Research in Child Development,* 1938, **3,** #6 pp. 1–254.

MacFarlane, J. W., Allen, L., & Honzik, M. P. A developmental study of the behavior problems of normal children between twenty-one months and fourteen years. *University of California Publications in Child Development,* 1954, **2.**

McArthur, C. Personalities of first and second children. *Psychiatry,* 1956, **19,** 47–54.

McCandless, B. R. *Children behavior and development.* New York: Holt, Rinehart & Winston, 1967.

Masling, J., Weiss, L., & Rothschild, B. Relationships of oral imagery to yielding behavior and birth order. *Journal of Consulting and Clinical Psychology,* 1968, Vol. 32, 89–91.

Mason, S. T. *A history of the sciences.* New York: Abelard-Schumann (Collier Books), 1962.

Mead, G. R. *Mind, self, and society.* Chicago: University of Chicago Press, 1934.

Miller, N. E. & Dollard, J. Social learning and imitation. New Haven: Yale Univ. Press. 1941.

Miller, N., & Zimbardo, D. G. Similarity vs. emotional comparison as motives for affiliation. Paper presented at the American Psychological Association, Chicago, May 1965.

Miller, N., & Zimbardo, D. G. Motives for fear-induced affiliation: Emotional comparison or interpersonal similarity. *Journal of Personality,* 1966, **341,** 481–503.

Millis, J., & Seng, Y. P. The effect of age and parity of the mother on birthweight of the offspring. *Annals of Human Genetics,* 1954, **19,** 58–73.

Moore, O. K. Some puzzling aspects of social interaction. *Review of Metaphysics,* 1962, **15,** 409–433.

Moore, R. A., & Ramseur, F. A study of the background of 100 hospitalized veterans with alcoholism. *Quarterly Journal for the Study of Alcoholism,* 1969, **21,** 51–67.

Moore, R. K. Susceptibility to hypnosis and susceptibility to social influence. *Journal of Abnormal and Social Psychology,* 1964, **68,** 282–294.

Moss, H. A. Sex, age, and state as determinants of mother-infant interaction. *Merrill-Palmer Quarterly,* 1967, **13,** 19–36.

Murdoch, P. H. J. Birth order and age at marriage. *British Journal of Social and Clinical Psychology,* 1966, **5,** 24–29.

Murphy, L. B., Murphy, G. & Newcomb, T. *Experimental social psychology,* New York: Harper & Brothers, 1937. Pp. 348–363.

Murray, H. A., Editor. *Explorations in personality: A clinical and experimental study of 50 men of college age.* New York: Oxford University Press, 1938.

Nagy, I. B., & Framo, J. L. *Intensive family therapy.* New York: Harper & Row, 1965.

Nash, J. The father in contemporary culture and current psychological literature. *Child Development,* 1965, **36**, 261–297.

Navratil, L. On the etiology of alcoholism. *Quarterly Journal for the Study of Alcoholism,* 1959, **20**, 236–244.

Neisser, E. G. *Brothers and sisters.* New York: Harper & Brothers, 1951.

Newfield, W. F. The effect of social desirability and birth order on the making of physical disabilities. Paper presented at the American Psychological Association, Washington, September 1966.

Nisbett, R. E. Birth order and participation in dangerous sports. *Journal of Personality and Social Psychology,* 1968, Vol. 8, No. 4, 351–353.

Nisbett, R. E., & Schachter, S. Cognitive manipulation of pain. *Journal of Experimental Social Psychology,* 1966, **2**, 227–236.

Nowicki, S. Birth order and personality: Some unexpected findings. *Psychological Reports,* 1967, **21**, 265–267.

Oberlander, M., & Jenkins, N. Birth order and academic achievement. *Journal of Individual Psychology,* 1967, **23**, 103–109.

Orbison, M. E. Some effects of parental maladjustment on first-born children. *Smith College Studies in Social Work,* 1945, **16**, 138–139.

Orlansky, H. Infant care and personality. *Psychological Bulletin,* 1949, **46**, 1–48.

Ostrovsky, E. S. *Children without men.* New York: Collier Books, 1959.

Palmer, R. D. Birth order and identification. *Journal of Consulting Psychology,* 1966, **30**, 129–135.

Parsely, M. The delinquent girl in Chicago: The influence of ordinal position and size of family. *Smith College Studies in Social Work,* 1933, **3**, 274–83.

Parsons, T., & Bales, R. F. *Family, socialization, and interaction process.* Glencoe, Ill.: Free Press, 1955.

Pascal, G. R., & Jenkins, W. O. A study of the early environment of workhouse inmate alcoholics and its relationship to adult behavior. *Quarterly Journal for the Study of Alcoholism,* 1960, **21**, 40–50.

Patterson, R., & Zeigler, T. W. Ordinal position and schizophrenia. *American Journal of Psychiatry,* 1941, **98**, 455–456.

Pepitone, A. *Attraction and hostility.* New York: Atherton Press, 1964.

Phelps, H. R., & Horrocks, J. E. Factors influencing informal groups of adolescents. *Child Development,* 1958, **29**, 69–86.

Phillips, E. L. Cultural vs. interpsychic factors in childhood. *Journal of Clinical Psychiatry,* 1956, **12**, 400–401.

Pierce, J. V. *The educational motivation of superior children who do and do not achieve in high school.* Washington, D.C.: U.S. Office of Education, Department of Health, Education, and Welfare, November 1959.

Plank, R. The family constellation of a group of schizophrenic patients. *American Journal of Orthopsychiatry*, 1953, **23**, 817–825.

Popper, K. *The poverty of historicism*. New York: Harper & Row, 1960.

Radke-Yarrow, M. The elusive evidence. Paper presented at the American Psychological Association, Philadelphia, September 1963.

Radloff, R. Opinion evaluation and affiliation. *Journal of Abnormal and Social Psychology*, 1961, **62**, 578–585.

Rhine, W. R. Birth order differences in conformity and level of achievement arousal. *Child Development*, 1968, **39**, 987–996.

Rhine, W. R. Birth order differences in resistance to conformity pressure related to social class and level of achievement arousal. *Proceedings of 77th Annual Convention*, A.P.A. 1969, 265–266.

Ring, K., Lipinski, C. E., & Braginsky, D. The relationships of birth order to self-evaluation, anxiety-reduction and susceptibility to emotional contagion. *Psychological Monographs*, 1965, **79**(Whole No. 603).

Roberts, J. M., and Sutton-Smith, B. Child training and game involvement, *Ethnology*, 1962, Vol. I, No. 2, 166–185.

Roberts, J. M., and Sutton-Smith, B. Cross-cultural correlates of games of chance, *Behavior Science Notes*, 1966, **3**, 131–144.

Roe, A. *The making of a scientist*. New York: Dodd, Mead, 1953.

Rosen, B. C. Family structure and achievement motivation. *American Sociological Review*, 1961, **26**, 574–585.

Rosen, B. C. Family structure and value transmission. *Merrill-Palmer Quarterly*, 1964, **10**, 59–76.

Rosenbaum, M. Psychological effects on the child raised by an older sibling. *American Journal of Orthopsychiatry*, 1963, **33**, 515–520.

Rosenberg, B. G., Goldman, R., & Sutton-Smith, B. Sibling age spacing effects on cognitive activity in children. Proceedings, 77th Annual Convention, A.P.A., 1969, 261–262.

Rosenberg, B. G., & Sutton-Smith, B. The measurement of masculinity and femininity in children. *Child Development*, 1959, **30**, 373–380.

Rosenberg, B. G., & Sutton-Smith, B. A revised conception of masculine-feminine differences in play activities. *Journal of Genetic Psychology*, 1960, **96**, 165–170.

Rosenberg, B. G., & Sutton-Smith, B. The measurement of masculinity and femininity in children: An extension and revalidation. *Journal of Genetic Psychology*, 1964, **104**, 259–264. (a)

Rosenberg, B. G., & Sutton-Smith, B. Ordinal position and sex role identification. *Genetic Psychological Monographs*, 1964, **70**, 297–328. (b)

Rosenberg, B. G., & Sutton-Smith, B. The relationship of ordinal position and sibling sex status to cognitive abilities. *Psychonomic Science*, 1964, **1**, 81–82. (c)

Rosenberg, B. G., & Sutton-Smith, B. Sibling association, family size, and cognitive abilities. *Journal of Genetic Psychology*, 1966, **109**, 271–279.

Rosenberg, B. G., & Sutton-Smith, B. Family interaction effects on masculinity-femininity. Paper presented at the Society for Research in Child Development, New York, March 1967.

Rosenberg, B. G., & Sutton-Smith, B. Family interaction effects on masculinity-femininity. *Journal of Personality and Social Psychology,* 1968, **8,** 117–120.

Rosenberg, B. G., & Sutton-Smith, B. Sibling age spacing effects on cognition. *Developmental Psychology,* 1969, **1,** 661–668.

Rosenberg, B. G., Sutton-Smith, B. & Griffiths, J. Sibling differences in empathic style. *Perceptual and Motor Skills,* 1965, **21,** 811–814.

Rosenberg, B. G., Sutton-Smith, B., & Landy, F. S. The effects of limited father absence on the cognitive and emotional development of children. Midwestern Psychological Association, May 1967.

Rosenberg, B. G., Sutton-Smith, B., & Morgan E. The use of opposite sex scales as a measure of psychosexual deviancy. *Journal of Consulting Psychology,* 1961, **25,** 221–225.

Rosenberg, M. *Society and the adolescent self-image.* Princeton: Princeton Univer. Press, 1965.

Rosenfeld, H. Relationships of ordinal position to affiliation and achievement motives: Direction and generality. Unpublished Manuscript, University of Kansas, 1966.

Rosenow, C. The incidence of first-born among problem children. *Journal of Genetic Psychology,* 1930, **37,** 145–151.

Rosenow, C., & Whyte, A. H. The ordinal position of problem children. *American Journal of Orthopsychiatry,* 1931, **1,** 430–434.

Rosenthal, D. Familial concordance by sex with respect to schizophrenia. *Psychological Bulletin,* 1966, **59,** 401–421.

Rosenthal, R. The effect of the experimenter on the results of psychological research. In B. Maher (Ed.), *Progress in experimental personality research.* Vol. I. New York: Academic Press, 1964.

Ross, B. M. Some traits associated with sibling jealousy in problem children. *Smith College Studies of Social Work,* 1931, **1,** 363–378.

Rothbart, M. L. K. *Birth order and mother-child interaction.* (Doctoral dissertation, Ph.D., Stanford University) Ann Arbor, Mich.: University Microfilms, 1967. No. 67–7961.

Salber, E. J. The effect of sex, birthrank, and birthweight on growth in the first year of life. *Human Biology,* 1957, **29,** 194–213.

Sampson, E. E. Birth order, need achievement and conformity. *Journal of Abnormal and Social Psychology,* 1962, **64,** 155–159.

Sampson, E. E. The study of ordinal position: Antecedents and outcomes. In B. Maher (Ed.), *Progress in experimental personality research.* New York: Academic Press, 1965.

Sampson, E. E., & Hancock, F. T. An examination of the relationship between ordinal position, personality, and conformity. *Journal of Personality and Social Psychology,* 1967, **5,** 398–407.

Sarbin, T. R., & Hardyck, C. Conformance in role perception as a personality variable. *Journal of Consulting Psychology,* 1955, **19,** 109–111.

Sarnoff, I., & Zimbardo, P. Anxiety, fear and social affiliation. *Journal of Abnormal and Social Psychology,* 1961, **62,** 155–159.

Schachter, S. Deviation, rejection and communication. *Journal of Abnormal and Social Psychology*, 1951, **44**, 190–207.

Schachter, S. *The psychology of affiliation*. Stanford, Calif.: Stanford University. Press, 1959.

Schachter, S. Birth order, eminence and higher education. *American Sociological Review*, 1963, **28**, 757–767.

Schachter, S. Birth order and sociometric choice. *Journal of Abnormal and Social Psychology*, 1964, **68**, 453–456.

Scharl, J. (Ed.) *Grimm's fairy tales*. London: Routledge, Kegan and Paul, 1948.

Schmuck, R. Sex of sibling, birth order position, and female disposition to conformity in two-child families. *Child Development*, 1963, **34**, 913–918.

Schooler, C. Birth order and schizophrenia. *Archives of General Psychiatry*, 1961, **4**, 91–97.

Schooler, C. Birth order and hospitalization for schizophrenia. *Journal of Abnormal and Social Psychology*, 1964, **69**, 574–579.

Schooler, C., & Caudill, W. Symptomatology in Japanese and American schizophrenics. *Ethnology*, 1964, **2**, 172–178.

Schooler, C., & Parkel, D. The overt behavior of chronic schizophrenics and its relationship to their internal state and personal history. *Psychiatry*, 1966, **29**(1), 67–77.

Schooler, C., & Raynsford, S. W. Affiliation among chronic schizophrenics: Relations to intrapersonal and background factors. *American Psychologist*, 1961, **14**, 358.

Schooler, C., & Scarr, S. Affiliation among chronic schizophrenics: Relation to intrapersonal and birth order factors. *Journal of Personality*, 1962, **30**, 178–192.

Schoonover, S. M. The relationship of intelligence and achievement to birth order, sex of sibling and age interval. *Journal of Educational Psychology*, 1959, **50**, 143–146.

Schultz, D. P. Birth order of volunteers for sensory restriction research. *Journal of Social Psychology*, 1967, **73**, 71–73.

Scott, J. P. *Animal behavior*. New York: Dover, 1963.

Sears, P. S. Doll play aggression in normal young children: Influence of sex and sibling status, father's absence. *Psychological Monographs*, 1951, **65**(Whole No. 323).

Sears, R. R. Ordinal position in the family as a psychological variable. *American Sociological Review*, 1950, **15**, 397–401.

Sears, R. R., Maccoby, E., & Levin, H. *Patterns of child rearing*. Evanston, Ill.: Row, Peterson, 1957.

Sears, R. R., Whiting, J. W. M., Nowlis, V., & Sears, P. S. Some child-rearing antecedents of aggression and dependency in young children. *Genetic Psychological Monographs*, 1953, **47**, 135–236.

Sechrest, L., & Flores, L. Sibling position of Philippine psychiatric patients. *Journal of Social Psychology,* 1969, **77,** 135–137.

Sells, S. B., & Roff, N. Peer acceptance-rejection and birth order. *American Psychologist,* 1963, **18,** 355.

Sewell, M. Two studies in sibling rivalry. I. Some causes of jealousy in young children. *Smith College Studies of Social Work,* 1930, **1,** 6–22.

Sheldon, P. M. The families of highly gifted children. *Journal of Marriage and Family Living,* 1954, **16,** 59–60.

Shrader, W. K., & Leventhal, T. Birth order of children and parental report of problems. *Child Development,* 1968, **39,** No. 4, 1165–1175.

Singer, D. *Creativity and sex role development,* Colloquium presentation Psychology Dept., Teachers College, Columbia University, November 1968.

Singer, J. E. The use of manipulative strategies: Machiavellianism and attractiveness. *Sociometry,* 1964, **27,** 128–150.

Singer, J. E., & Lamb, P. F. Social concern, body size, and birth order. *Journal of Social Psychology,* 1966, **68,** 143–151.

Singer, J. E., & Shockley, V. L. Ability and affiliation. *Journal of Personality and Social Psychology,* 1965, **1,** 95–100.

Singer, J. L. *Daydreaming.* New York: Random House, 1966.

Sinha, Jai B. P. Birth order and sex differences in n-achievement and n-affiliation. *Journal of Psychological Researches,* 1967, **11,** 22–27.

Skinner, G. W. *Filial sons and their sisters: Configuration and culture in Chinese families.* Paper presented at the conference "Kinship in Chinese Society," Social Science Research Council. New York. 1966.

Slawson, J. *The delinquent boy: A socio-psychological study.* Boston: Badger, 1926.

Sletto, R. F. Sibling position and juvenile delinquency. *American Journal of Sociology,* 1934, **39,** 657–669.

Smalley, R. E. Two studies in sibling rivalry. II. The influence of differences in age, sex and intelligence in determining the attitudes of siblings toward each other. *Smith College Studies of Social Work,* 1930, I, 23–40.

Smart, R. G. Alcoholism, birth order, and family size. *Journal of Abnormal and Social Psychology,* 1963, **66,** 17–23.

Smart, R. G. Social-group membership, leadership, and birth order. *Journal of Social Psychology,* 1965, **67,** 221–225.

Solomon, D. Birth order, family composition and teaching style. *Psychological Reports,* 1965, **17,** 871–874.

Solomon, L., & Nuttall, R. Sibling order, premorbid adjustment and remission in schizophrenia. *Journal of Nervous and Mental Disease,* 1967, **144,** 37–46.

Solomons, G., & Solomons, H. C. Factors affecting motor performances in four-month-old infants. *Child Development,* 1964, **35,** 1283–1296.

Spiegel, J. P. The resolution of role conflict within the family. *Psychiatry,* 1957, **20,** 1–16.

Spock, B. *Baby and child care.* New York: Cardinal Pocketbooks, 1957.

Stagner, R., & Katzoff, E. T. Personality as related to birth order and family size. *Journal of Applied Psychology,* 1963, **20**, 340–346.

Staples, F. R., & Walters, R. H. Anxiety, birth order and susceptibility to social influence. *Journal of Abnormal and Social Psychology,* 1961, **62**, 716–719.

Stendler, C. B. Possible causes of over-dependency in young children. *Child Development,* 1954, **25**, 125–147.

Stevenson, H. W. Is the human personality more plastic in infancy and childhood? *American Journal of Psychiatry,* 1957, **114**, 152–161.

Stevenson, H. W. Social reinforcement of children's behavior. In L. P. Lipsett and C. C. Spiker (Eds.), *Advances in child development and behavior.* New York: Academic Press, 1965, Volume II. Pp. 97–126.

Steward, R. H. Birth order and dependency. *Journal of Personality and Social Psychology,* 1967, **6**, 192–194.

Storer, N. W. Ordinal position and the Oedipus complex. *Laboratory of Social Relations of Harvard University Bulletin,* 1961, **10**, 18–21.

Stotland, E. Social schemas and birth order. Symposium presented at the meeting of the American Psychological Association, San Francisco, September 1968.

Stotland, E., & Cottrell, N. B. Similarity of performance as influenced by interaction, self-esteem, and birth order. *Journal of Abnormal and Social Psychology,* 1962, **64**, 183–191.

Stotland, E., & Dunn, R. E. Identification, opposition, authority, self-esteem and birth order. *Psychological Monographs,* 1962, **76** (Whole No. 528).

Stotland, E., & Dunn, R. E. Empathy, self-esteem, and birth order. *Journal of Abnormal and Social Psychology,* 1963, **66**, 532–540.

Stotland, E., & Walsh, J. A. Birth order and an experimental study of empathy. *Journal of Abnormal and Social Psychology,* 1963, **66**, 610–614.

Stout, A. M. *Parent behavior toward children of differing ordinal position and sibling status.* Unpublished Ph.D. dissertation, University of California, Berkeley, 1960.

Stratton, G. M. Anger and fear: Their probable relation to each other, to intellectual work and to primogeniture. *American Journal of Psychology,* 1927, **39**, 125–140.

Stratton, G. M. The relation of emotion to sex primogeniture and disease. *American Journal of Psychology,* 1934, **46**, 590–595.

Strauss, B. V. The dynamics of ordinal position effects. *Quarterly Journal of Child Behavior,* 1951, **3**, 133–145.

Strodtbeck, F. L. The family as a three-person group. *American Sociological Review,* 1954, **19**, 23–29.

Strodtbeck, F. L., & Creelan, P. Interaction, linkage between family size, intelligence, and sex role identity. *Journal of Marriage and Family,* 1968, **30**, 301–307.

Stroup, A. L., & Hunter, K. J. Sibling position in the family and personality of offspring. *Journal of Marriage and the Family*, 1965, **27**, 65–68.

Stuart, J. C. Data on the alleged psychopathology of the only child. *Journal of Abnormal and Social Psychology*, 1926, **20**, 441.

Suedfeld, P. Birth order of volunteers for sensory deprivation. *Journal of Abnormal and Social Psychology*, 1964, **68**, 195–196.

Suedfeld, P. Anticipated and experienced stress in sensory deprivation as a function of orientation and ordinal position. *Journal of Social Psychology*, 1968, **76**, 259–263.

Suedfeld, P. Sensory deprivation stress: Birth order and instructional set as interacting variables. *Journal of Personality and Social Psychology*, 1969, **11**, 70–74.

Sundararaj, N., & Sridhora, R. Order of birth and schizophrenia. *British Journal of Psychiatry*, 1966, **112**, 1127–1129.

Sutton-Smith, B. Play preference and play behavior: A validity study. *Psychological Reports*, 1965, **16**, 65–66.

Sutton-Smith, B. Developmental laws and the experimentalists' ontology. Symposium presented at the American Psychological Association, Washington, D.C., September 1966. (a)

Sutton-Smith, B. Role replication and reversal in play. *Merrill-Palmer Quarterly*, 1966, **22**, 993–994. (b)

Sutton-Smith, B. The value of dramatic role-playing: An experimental confirmation. *Independent Schools Association of the Central States Bulletin*, 1966, **5**, 26–27. (c)

Sutton-Smith, B. Modelling and reaction in sibling interaction. Paper presented to Minnesota Symposium on Child Development, University of Minnesota, May 17, 1968.

Sutton-Smith, B., & Roberts, J. M. Game involvement in adults, *Journal of Social Psychology*, 1963, **60**, 15–30.

Sutton-Smith, B. & Roberts, J. M. Rubrics of competitive behavior. *Journal of Genetic Psychology*, 1964, **105**, 13–37.

Sutton-Smith, B., & Roberts, J. M. Studies in an elementary strategic game. *Genetic Psychological Monographs*, 1967, **75**, 3–42.

Sutton-Smith, B., Roberts, J. M., & Rosenberg, B. G. Sibling association and role involvement. *Merrill-Palmer Quarterly*, 1964, **10**, 25–38.

Sutton-Smith, B., & Rosenberg, B. G. A scale to identify impulsive behavior in children. *Journal of Genetic Psychology*, 1959, **95**, 211–216.

Sutton-Smith, B., & Rosenberg, B. G. Manifest anxiety and games preferences in children. *Child Development*, 1960, **31**, 307–311.

Sutton-Smith, B., & Rosenberg, B. G. Impulsivity and peer perception. *Merrill-Palmer Quarterly*, 1961, **7**, 233–238. (a)

Sutton-Smith, B., & Rosenberg, B. G. Impulsivity and sex preference. *Journal of Genetic Psychology*, 1961, **98**, 187–192. (b)

Sutton-Smith, B., & Rosenberg, B. G. Age changes in the effects of ordinal position on sex role identification. *Journal of Genetic Psychology*, 1965, **107**, 61–73. (a)

Sutton-Smith, B., & Rosenberg, B. G. Sibling perception of power styles within the family. Paper presented at the American Psychological Association, Chicago, September 1965. (b)

Sutton-Smith, B., & Rosenberg, B. G. The dramatic sibling. *Psychological Report*, 1966, **22**, 993–994. (a)

Sutton-Smith, B., & Rosenberg, B. G. *A factor analysis of power styles in the family*. Paper presented at the American Psychological Association, New York, September 1966. (b)

Sutton-Smith, B. & Rosenberg, B. G. The dramatic boy. *Perceptual and Motor Skills*, 1967, **25**, 247–248.

Sutton-Smith, B., & Rosenberg, B. G. Sibling consensus on power tactics. *Journal of Genetic Psychology*, 1968, **112**, 63–72.

Sutton-Smith, B., Rosenberg, B. G., & Houston, S. Sibling perception of parental models. Paper presented at the Eastern Psychological Association, Washington, D.C., April 20, 1968.

Sutton-Smith, B., Rosenberg, B. G., & Landy, F. The interaction of father absence and sibling presence on cognitive abilities, *Child Development*, 1968, Vol. 39, No. 4, 1213–1221.

Sutton-Smith, B., Rosenberg, B. G., & Morgan, E. Historical changes in the freedom with which children express themselves on personality inventories. *Journal of Genetic Psychology*, 1961, **99**, 309–315.

Sutton-Smith, B., Rosenberg, B. G., & Morgan, E. The development of sex differences in play choices during preadolescence. *Child Development*, 1963, **34**, 119–126.

Szasz, T. S. *The myth of mental illness*. New York: Hoeber-Harper, 1964.

Taylor, L. The social adjustment of the only child. *American Journal of Sociology*, 1945, **51**, 227–232.

Taylor, R. E., & Eisenman, R. Birth order and sex differences in complexity-simplicity. Color-form preference and personality. *Journal of Projective Techniques and Personality Assessment*, 1968, **32**, No. 4, 383–387.

Teepen, N. S. Sibling relationships in sex role identification. M.A. Thesis, Ohio State University, 1963.

Terman, L. M. *Genetic studies of genius*. Palo Alto: Stanford Univ., 1925.

Thelen, M. H. The relationship of selected variables to intra-family similarity of defense preferences. *Journal of Projective Techniques and Personality Assessment*, 1967, **31**, 23–28.

Thornton, D. A., & Arrowood, A. J. Self-evaluation, self-enhancement, and the locus of social comparison. *Journal of Experimental Social Psychology*, 1966, **1**, 40–48.

Thurstone, L. L., & Jenkins, R. L. Birth order and intelligence. *Journal of Educational Psychology*, 1929, **20**, 640–651.

Thurstone, L. L., & Jenkins, R. L. *Order of birth, parentage and intelligence*. Chicago: University of Chicago Press, 1931.

Toman, W. *Family constellation: Theory and practice of a psychological game*. New York: Springer, 1961.

Toman, W., & Gray, B. Family constellations of "normal" and "disturbed" marriages: An empirical study. *Journal of Individual Psychology*, 1961, **17**, 93–95.

Torrance, E. G. A psychological study of American jet aces. Paper read at the Western Psychological Association, Long Beach, California, 1954.

Tsuang, Ming- Tso. Birth order and maternal age of psychiatric in-patients. *British Journal of Psychiatry*, 1966, **112**, 1131–1141.

Tuckman, J., & Regan, R. A. Ordinal position and behavior problems in children. *Journal of Health and Social Behavior*, 1967, **8**, 32–39.

U.S. Department of Health, Education, and Welfare. *Vital statistics of the United States, Vol. I: Natality*. Washington, D.C., 1961.

Varela, J. A. A cross-cultural replication of an experiment involving birth order. *Journal of Abnormal and Social Psychology*, 1964, **69**, 456–457.

Veroff, J. Development and validation of a projective measure of power motivation. *Journal of Abnormal and Social Psychology*, 1957, **54**, 1–8.

Very, P. S. & Zannini, J. A. Relation between birth order and being a beautician. *Journal of Applied Psychology*, 1969, **53**, 149–151.

Visher, S. S. Environmental backgrounds of leading American scientists. *American Sociological Review*, 1948, **13**, 66–72.

Vogel, W., & Lauterbach, C. G. Sibling patterns and social adjustment among normal and psychiatrically disturbed soldiers. *Journal of Consulting Psychology*, 1963, **27**, 236–242.

Vroegh, K. The relationship of birth order and sex of sibling to masculinity and femininity. Paper presented at the Society for Research in Child Development, Santa Monica, March 1969.

Wagner, N. N. Birth order of volunteers: Cross-cultural data. *Journal of Social Psychology*, 1968, **74**, 133–134.

Waldrop, M. F. Effects of family size and density on newborn characteristics. *Journal of Orthopsychiatry*, 1965, **35**, 342–343.

Walker, C. E., & Tahmisian, J. Birth order and student characteristics. *Journal of Consulting Psychology*, 1967, **31**, 219.

Walker, C. E., & Tahmisian, J. Birth order and student characteristics: A replication. *Journal of Consulting Psychology*, 1967, **31**, 219.

Walters, R. H., & Ray, E. Anxiety, isolation and reinforcer effectiveness. *Journal of Personality*, 1960, **28**, 358–367.

Ward, C. S. A further examination of birth order as a selective factor among volunteer subjects. *Journal of Abnormal and Social Psychology*, 1964, **69**, 311–313.

Warren, J. R. Birth order and social behavior. *Psychological Bulletin*, 1966, **65**, 38–49.

Weinstein, L. Social experience and social schemata. *Journal of Personality and Social Psychology*, 1967, **6**, 429–434.

Weiss, J. H., Wolff, A., & Wistley, R. Birth order, recruitment conditions, and preference for participation in group vs. non-group experiments. *American Psychologist*, 1963, **18**, 356.

Weiss, J. H. Birth order and physiological stress response. *American Psychologist*, 1967, **22**, 557. (a)

Weiss, J. H. Correlates of birth order in normal and asthmatic children. Paper presented at the American Psychological Association, Washington, D.C., 1967. (b)

Weiss, R. L. Acquiescence response set and birth order. *Journal of Consulting Psychology*, 1966, **30**, 365.

Weller, G. M. Arousal effects on the newborn infant of being first or later born. *Journal of Orthopsychiatry*, 1965, **35**, 341–342.

Weller, G. M., & Bell, R. Q. Basal skin conductance and neonatal state. *Child Development*, 1965, **36**, 647–657

Weller, L. The relationship of birth order to anxiety. *Sociometry*, 1962, **25**, 415–417.

West, S. S. Sibling configuration of scientists. *American Journal of Sociology*, 1960, **66**, 268–274.

Wheeler, L. Motivation as determinant of upward comparison. *Journal of Experimental Social Psychology*, 1966, **1**, 27–31.

Whiting, J., & Child, I. *Child training and personality.* New Haven, Conn.: Yale Univer. Press, 1953.

Wile, I. S., & Davis, R. The relation of birth to behavior. *American Journal of Orthopsychiatry*, 1941, **11**, 320–334.

Wile, I. S., & Jones, A. B. Ordinal position and the behavior disorders of young children. *Journal of Genetic Psychology*, 1937, **51**, 61–63.

Wile, I. S., & Noetzel, E. A study of birth order and behavior. *Journal of Social Psychology*, 1931, **2**, 52–71.

Willis, C. B. The effects of primogeniture on intellectual capacity. *Journal of Abnormal and Social Psychology*, 1924, **18**, 375–377.

Willis, R. H., & Hollander, E. P. An experimental study of three response modes in social influence situations. *Journal of Abnormal and Social Psychology*, 1964, **69**, 150–156.

Wilson, P. R., & Patterson, J. Sex differences in volunteering behavior. *Psychological Reports*, 1965, **16**, 976.

Winch, R. *The familial determinants of identification.* Indianapolis: Bobbs-Merrill, 1962.

Winterbottom, M. R. The relation of need for achievement to learning experiences in independence and mastery. In J. W. Atkinson (Ed.), *Motives in fantasy, action, and society.* Princeton, N.J.: Van Nostrand, 1958, Pp. 453–476.

Witty, P. A. Only and intermediate children of high school ages. *Psychological Bulletin*, 1934, **31**, 734

Wohlford, P., & Jones, M. R. Ordinal position, age, anxiety and defensiveness in unwed mothers. Proceedings, 75th Annual Convention, American Psychological Association, Washington, D.C., 1967.

Wolf, A. Recruitment pressure, birth order, and sex as possible sources of bias in the recruitment of volunteer subjects. Paper presented at the Eastern Psychological Association, 1966.

Wolf, A., & Wolf, G. First born and last born: A critique. *Journal of Individual Psychology*, 1965, **21**(2), 159–164.

Wolf, K. A comparison of sibling position and academic achievement in elementary school. *Dissertation Abstracts,* 1967, Vol. 28.

Wolkon, G. H. *A comment on "alcoholism, birth order and family size."* Report from Mental Health Rehabilitation Research (Hill House), Cleveland, 1964.

Wolkon, G. H. Birth order, desire for and participation in psychiatric posthospital services. *Journal of Consulting and Clinical Psychology,* 1968, **32,** 42–46.

Wolkon, G. H., & Levinger, G. Birth order and need for achievement. *Psychological Reports,* 1965, **16,** No. 1, 73–74.

anxiety. *Journal of Abnormal and Social Psychology,* 1960, **61,** 216–222.

Wuebben, P. L. *Honesty of subjects· and birth order.* Abstract of doctoral dissertation, University of Wisconsin, 1967.

Yarrow, L. (Ed.) Symposium on personality consistency and change: Perspec-

Wrightsman, L. S. Effects of waiting with others on changes in level of felt tives from longitudinal research. *Vita Humana,* 1964, **2,** 7.

Yaryan, R. B., & Festinger, L. Preparatory action and belief in the probable occurrence of future events. *Journal of Abnormal and Social Psychology,* 1961, **63,** 603–606.

Yasuda, S. A methodological inquiry into social mobility. *American Sociological Review,* 1964, **29,** 16–23.

Yoda, A., & Fukatsu, A. Ordinal position, birth order and personality. *Japanese Journal of Educational Psychology,* 1963, **11,** 239–246.

Yourglich, A., Explorations in the sociological study of sibling systems. *Family Life Coordinator,* 1964, **13,** 91–95.

Yourglich, A., & Schiessl, D. Constructiong a sibling systems measurement device. *Family Life Coordinator,* 1966, **15,** 107–111.

Ziegler, F. J., Imboden, J. B., & Meyer, E. Contemporary conversion reaction. *American Journal of Psychiatry,* 1960, **116,** 903–908.

Zimbardo, P., & Formica, R. Emotional comparison and self-esteem as determinants of affiliation. *Journal of Personality,* 1963, **31,** 141–162.

Zucker, R., Manoswitz, M., & Lanyon, R. Birth order, anxiety and affiliation during a crisis. *Journal of Personality and Social Psychology,* 1968, Vol. 8, No. 4, 354–359.

Name Index

Subject Index

191